POLITICS AND EXPERTISE

Politics and Expertise

HOW TO USE SCIENCE IN A DEMOCRATIC SOCIETY

ZEYNEP PAMUK

PRINCETON UNIVERSITY PRESS
PRINCETON & OXFORD

Published by Princeton University Press
41 William Street, Princeton, New Jersey 08540
99 Banbury Road, Oxford OX2 6JX

press.princeton.edu

All Rights Reserved

Library of Congress Control Number: 2021944419

First paperback printing, 2024
Paper ISBN 9780691219264
Cloth ISBN 9780691218939
ISBN (e-book) 9780691218946

British Library Cataloging-in-Publication Data is available

Editorial: Matt Rohal
Production Editorial: Nathan Carr
Jacket/Cover Design: Karl Spurzem
Production: Erin Suydam
Publicity: Alyssa Sanford and Charlotte Coyne
Copyeditor: Cindy Milstein

Jacket/Cover Credit: M. Style / Shutterstock

This book has been composed in Arno

To my parents

It would not only be foolish, *but downright irresponsible* to accept the judgment of scientists and physicians without further examination.

—PAUL FEYERABEND, *SCIENCE IN A FREE SOCIETY*

The rational layman will recognize that, in matters about which there is good reason to believe that there is expert opinion, he ought (methodologically) not to make up his own mind.

—JOHN HARDWIG, "EPISTEMIC DEPENDENCE"

CONTENTS

ACKNOWLEDGMENTS

THIS BOOK BEGAN as a dissertation project, and would not have been possible without the unfailing support and critical scrutiny of my dissertation committee. Nancy Rosenblum infused the project with her intellectual energy, asked sharp questions, and gave the soundest advice at crucial moments. Richard Tuck taught to me to think creatively, read widely, keep an open mind, and develop a nose for theoretical puzzles. Eric Beerbohm directed me to the most relevant literatures and provided critical suggestions. Dennis Thompson stepped out of retirement to lend his knowledge and experience to the project. They all believed in this project from the start, imagined what my ideas could become before I did, and encouraged me to follow my arguments to their conclusions, however radical. Their wisdom, kindness, and generosity remain my inspiration. I am also greatly indebted to Frances Kamm for teaching me how to think clearly, Peter Hall for introducing me to Thomas Kuhn, and Danielle Allen for invaluable advice on writing.

My ideas were much improved over the years by conversations with and suggestions from Jacob Abolafia, Adriana Alfaro Altamirano, Avishay Ben Sasson-Gordis, James Brandt, Jonathan Bruno, Greg Conti, Leah Downey, Michael Frazer, Jonathan Gould, Natalia Gutkowski, John Harpham, Madhav Khosla, Suzie Kim, Jack Knight, Tsin Yen Koh, Monica Magalhaes, Alison McQueen, Julie Miller, Lowry Pressly, Rob Reich, Michael Rosen, Jacob Roundtree, Wendy Salkin, Alicia Steinmetz, Aleksy Tarasenko-Struc, Don Tontiplaphol, Beth Truesdale, and George Yin.

As a postdoctoral fellow at Oxford, I found a welcoming political theory community and shelter from the uncertainties of the academic job market. I am deeply grateful to Cécile Laborde for mentorship, advice, and inspiration during this crucial period as I was finding my feet after graduate school. Sophie Smith and Amia Srinivasan offered the perfect combination of incisive feedback, personal advice, and friendship. Cécile Fabre read the whole manuscript and suggested a number of changes that significantly improved it. Many colleagues and friends at Oxford read parts of the book, offering valuable comments and questions. Special thanks are due to Samuel Bagg, Teresa Bejan, Udit Bhatia, Paul Billingham, Daniel Butt, Tae-Yeoun Keum, Nikolas Kirby, Joseph Lacey, Maxime Lepoutre, Daniel Luban, David Miller, Alex Prescott-Couch, Élise Rouméas, Jay Ruckelshaus, Zofia Stemplowska, Anthony Taylor, Jonathan Wolff, and Annette Zimmermann. I am grateful to Walter Mattli and Maggie Snowling for their personal support and for cultivating a supportive intellectual environment at St John's College. Desmond King, Melissa Lane, Annabelle Lever, Lois McNay, Philip Pettit, and Laura Valentini provided encouragement at critical junctures.

A manuscript workshop at St John's College improved the book immeasurably. I would like to thank Simone Chambers, Michael Frazer, Alfred Moore, Nicholas Southwood, Kai Spiekermann, and audience members for feedback that struck just the right balance between critical and constructive. It was humbling to see that scholars I barely knew were willing to take the time to engage with the whole manuscript.

This project was supported by the Harvard University Department of Government, Edmond J. Safra Center for Ethics, and St John's College, Oxford. I would like to thank them for their generous financial support and stimulating intellectual environments. Parts of this book were presented at the Duke and Princeton graduate conferences in political theory; the universities of Amsterdam, East Anglia, Leiden, Nottingham, and Sabancı; University College London; Stanford University; Blavatnik School of Government; Green Templeton College; Institute of Historical Research; Irish Philosophical Society; MANCEPT; Nuffield College; Oxford Centre for the Study of Social Justice; and Marquis Cornwallis pub in London. I would like to thank Uğur Aytaç, Michael Frazer, Adela Halo, Michael Hannon, Gürol Irzık, Faik Kurtulmuş, Cara Nine, Andrei Poama, David Ragazzoni, and Greg Whitfield for the invitations and warm hospitality, and the audiences for many questions and comments.

I completed this book as an assistant professor at the University of California, San Diego. The Department of Political Science and chair Thad Kousser were incredibly supportive and accommodating during my transition in the midst of the COVID-19 pandemic. I am grateful to Sean Ingham and David Wiens for invaluable suggestions on the whole manuscript, encouragement and support during the completion of the book, and great conversations on the methods and aims of political theory. Our biweekly Zoom meetings were a highlight of the pandemic period.

I am also grateful to two anonymous reviewers for insightful and constructive feedback. They understood the aims of the project and offered many suggestions to improve it on its own terms. I would like to thank Matt Rohal and Rob Tempio at Princeton University Press for their trust in the project, and competence and enthusiasm at every stage of the editorial process. I am delighted that this book joins their excellent selections in political theory. Thanks also to Cindy Milstein for their careful copyediting.

Parts of this book have appeared in "Justifying Public Funding for Science," *British Journal of Political Science* (2019) and "Risk and Fear: Restricting Science under Uncertainty," *Journal of Applied Philosophy* (2020). I thank the publishers for permission to use these materials.

A group of friends deserve special recognition for their unwavering support throughout this period. Soledad Artiz Prillaman, Madhav Khosla, Tsin Yen Koh, Zeynep Koray, Sarah Nutman, and Lale Tüzmen have proved that the closest friendships can be sustaining even across continents and time zones. This book bears their imprint in ways that are difficult to express.

Ranjit Lall has been there at every step of the way this past decade. Many of the ideas in this book were tested on him first. He is my closest intellectual companion, most trusted adviser, and best friend. Finally, my deepest gratitude is to my parents, Yeşim Arat and Şevket Pamuk. Their deep and unconditional trust in me has been my greatest source of strength, and their dinner table interrogations have continually challenged and motivated me. They are my models for how to live a gracious intellectual life. I dedicate this book to them with love.

POLITICS AND EXPERTISE

1

Science on Trial

ON OCTOBER 22, 2012, in the small Italian town of L'Aquila, seven earthquake experts were convicted of manslaughter and sentenced to six years in prison.[1] The prosecutor claimed that they were responsible for the death of 309 residents in a major earthquake in 2009 due to their failure to adequately assess and communicate seismic risks ahead of time. In the three months preceding the earthquake, the city had experienced an event that experts call a seismic swarm: two or three low-level tremors daily. An additional fifty-seven tremors took place in the five days before. Residents were unnerved and turned to scientists for guidance on whether these tremors signaled a major earthquake, and if so, whether they should evacuate the city. Their worries were exacerbated when a local lab technician named Giampaolo Giuliani began to predict a major earthquake on the basis of his measurement of radon gas levels.[2] The scientific community had repeatedly rejected the reliability of radon measurements for short-term predictions of earthquakes, and Giuliani had been denied funding for his research several times because his work was insufficiently scientific.[3] But this did not stop him from setting up a website to post daily radon readings and sharing his predictions with the locals. A few days before the earthquake, the mayor issued a gag order on Giuliani for fear that his website would provoke panic.

---

1. Elisabetta Polovedo and Henry Fountain, "Italy Orders Jail Terms for 7 Who Did Not Warn of Earthquake," *New York Times*, October 22, 2012, https://www.nytimes.com/2012/10/23/world/europe/italy-convicts-7-for-failure-to-warn-of-quake.html.

2. Stephen S. Hall, "Scientists on Trial: At Fault?," *Nature*, September 14, 2011, https://www.nature.com/news/2011/110914/full/477264a.html.

3. John Dollar, "The Man Who Predicted an Earthquake," *Guardian*, April 5, 2010, https://www.theguardian.com/world/2010/apr/05/laquila-earthquake-prediction-giampaolo-giuliani.

It was in this context that the Italian Civil Protection Department and local officials decided to hold a meeting with seven seismologists to comment on the probability that the seismic swarm in L'Aquila might precede a major earthquake. The scientific opinion was that this is quite rare. According to the meeting minutes, one of the participating scientists said, "It is unlikely that an earthquake like the one in 1703 could occur in the short term, but the possibility cannot be totally excluded."[4] The meeting was short and followed by a press conference in which Bernardo De Bernardinis, vice director of the Civil Protection Department, announced that the situation was "certainly normal," adding, "The scientific community tells me there is no danger because there is an ongoing discharge of energy."[5] This press conference was the grounds for the charges that led to the scientists' conviction. The charge was not a failure to predict the earthquake, which the prosecutor recognized was not possible, but rather the misleading assurance by a group of respected experts that there was no danger. He claimed that this message had led residents—and especially the younger and more educated ones—to change their plans and stay in L'Aquila, with disastrous consequences.[6]

This small but dramatic episode illustrates some of the key features of the use and misuse of scientific advice in public policy.[7] On the one hand, it shows the dependence of citizens and public officials on scientific expertise on a matter literally of life and death.[8] The residents of L'Aquila turned to science for an explanation in the face of an unusual and frightening natural event. The science was crucial on this issue. Attempting to see the problem merely as a conflict over values, such as whether the residents were the sorts of people

---

4. Hall, "Scientists on Trial."

5. Nicola Nosengo, "Italian Court Finds Seismologists Guilty of Manslaughter," *Nature*, October 22, 2012, https://www.nature.com/news/italian-court-finds-seismologists-guilty-of-manslaughter-1.11640.

6. Hall, "Scientists on Trial."

7. For a detailed discussion of the role of values and uncertainty in this case, see Melissa Lane, "When the Experts Are Uncertain: Scientific Knowledge and the Ethics of Democratic Judgment," *Episteme* 11, no. 1 (March 2014): 97–118.

8. Deborah Coen argues that historically, earthquake science relied heavily on data provided by local observers. This changed in the twentieth century as a result of what she calls the construction of incommensurability between lay experience and scientific data, but she suggests that the twenty-first century may see another reversal given the increased uncertainty. See Deborah Coen, *The Earthquake Observers: Disaster Science from Lisbon to Richter* (Chicago: University of Chicago Press, 2013).

who would leave their city when faced with an existential threat, would be to miss the point. Factual questions mattered: What was the likelihood of a major earthquake, and what was the risk of harm to the residents in the event of an earthquake?

On the other hand, the incident exposes the limits of decision-making on the basis of scientific knowledge. Like many other areas of science, though more so than most, earthquake science is uncertain and inexact. Scientists have become increasingly capable of predicting the likelihood that an earthquake will strike a given area within a given time period, but there is still no accepted scientific method for reliable short-term prediction.[9] The seismologists who were consulted had some data on the likelihood of a major earthquake in the days following a seismic swarm, but these findings were far from conclusive. Given the uncertainty and limits of reliable knowledge, residents' attitudes toward risk were critical to determining the appropriate earthquake response. Yet ironically, only the lab technician Giuliani seemed to appreciate the power of public fear, while local officials appealed to the authority of science in an ill-conceived attempt to reassure the public.

After the highly publicized trial, scientists and scientific associations around the world protested the conviction on the grounds that it penalized scientists for making a prediction that turned out to be incorrect. The president of the American Association for the Advancement of Science wrote a letter to the president of Italy, arguing that this kind of treatment would have a chilling effect and discourage scientists from public engagement. While the scapegoating of scientists through the criminal system may not have been an appropriate response to what had taken place, it was clearly a reaction to the mishandling of expert advice before the earthquake. The officials had denied the public a chance to understand the content and uncertainty of the science, instead delivering an authoritative judgment with an appeal to the views of "the scientific community." This had created a false sense of security, and deprived citizens of the ability to evaluate the information for themselves, and make up their own minds about how to respond to an unknown and unquantified danger.

The L'Aquila case was a particularly dramatic example of a community's dependence on scientific advice and the disastrous results of bad advice, but it is hardly unique. The COVID-19 pandemic, which started in Wuhan, China,

9. Polovedo and Fountain, "Italy Orders Jail Terms for 7 Who Did Not Warn of Earthquake."

in late 2019, and killed nearly two million people globally within a year, exposed both the dependence of governments on scientific advice and cracks in this relationship on a much greater scale. In the face of a new and catastrophic risk, the lives of billions depended on scientists' ability to study the behavior of the novel coronavirus, provide policy advice to governments, and produce safe and effective vaccines. Governments turned to scientists for help, and scientists delivered remarkable amounts of new knowledge in a short period of time. At the same time, this episode showed the difficulties of using scientific knowledge under conditions of uncertainty and disagreement—and the severe costs of failure. Many governments claimed to be following the science while pursuing wildly different policies. Scientists publicly disagreed among themselves as well as with government policies. The science itself was evolving rapidly. Key aspects of the disease, from transmission and fatality rates to the duration of immunity, were unknown. Scientists and public health officials who appeared on regular press conferences focused on short-term health objectives, while disregarding the economic and social impacts of policies as well as broader conceptions of health. Their assumptions were not always disclosed or scrutinized. As appeals to the authority of scientific models and findings dominated public discourse, rejections and dismissals of scientific authority from politicians and the public also intensified.

The COVID-19 response of many countries involved serious mistakes and with disastrous results. Social scientists, public health experts, and physicians are studying the effects of these policies and trying to explain why some nations fared better than others. It is difficult to diagnose the failure, however, without relying on an account that articulates the sources of tension in the relationship between science and democracy, and examines better and worse ways to mitigate them. This book seeks to offer such an account.

What are the dilemmas of scientific advisory committees and their proper role within broader democratic decision-making procedures? How should the certainty, reliability, and completeness of available scientific knowledge affect the procedures for its use? Is it appropriate to expect citizens to engage with the technicalities of science? How are questions about the use of science in a democratic society influenced by broader decisions about the funding, design, and conduct of scientific research? These are the questions I set out to answer. The answers, in turn, will help us identify the structural tensions in the science-democracy relationship, and distinguish them from contingent problems due to the moral failings or incompetence of individuals occupying prominent political or scientific positions at a particular time.

———

Our ability to act on some of the biggest problems of our times, such as pandemics, climate change, biotechnology, nuclear weapons, or environmental issues, requires relying on knowledge provided by scientists and other experts. The modern state has struck an unprecedented partnership with science, taking scientific inquiry as its authoritative source of knowledge and the means for bringing about better policy outcomes. New scientific research determines what we see as our problems and the range of options we have for solving them. Meanwhile, contemporary political life is increasingly characterized by pathological treatments of expertise, with denials of science and distrust of scientists, on the one hand, and appeals to the authority of experts and complaints about the ignorance of the citizenry, on the other. These attitudes are intensified in reaction to one another: frustration with denial and pseudoscience leads to increased appeals to the authority of scientists, which in turn generates resentment—and more denial. It is a vicious cycle.

The partnership between democracy and expertise is intrinsically unstable. Democracy—rule by the people—holds out the promise that the people can shape their collective life by making decisions together, either directly or through elected representatives. Expert knowledge threatens to alter or limit the possibilities for democratic decision-making. It presents a rival source of authority in the public sphere, based on truth rather than agreement. This creates the danger that the authority of experts and their claims to objective knowledge will crowd out the space for democratic judgment about how to shape a collective existence. At the same time, scientific experts have no direct access to political power. The truth of scientific claims may not depend on the number of people who believe in them, but their uptake in politics inevitably requires persuading the many. In the realm of politics, scientists must appeal to people who do not share and may not understand the scientific community's methods for settling the truth. Citizens and their representatives ultimately retain the right to reject scientific knowledge, which is a right that they exercise quite often.

Efforts to eliminate this inherent tension would be problematic for both science and politics. Determining scientific truth democratically would be irrational and dangerous, while justifying democratic decisions by appeal to standards of scientific correctness would set a standard both impossibly high and inappropriate for politics. The legitimacy of democratic decisions derives not from their scientific credentials but instead from the fact that those who are subjected to them have had a say in the decision process.[10] The challenge, then, is to devise ways for expertise and democracy to coexist productively. Expert knowledge could be used to expand the power of democracy or lead to the alienation of citizens from a politics that seems to defy their control. The success of the relationship between democracy and expertise depends on whether democracies can find ways to use expertise to further their own ends and produce good outcomes. Recent failures in the use of science for political decisions—not only on COVID-19, but on climate change, vaccines, genetically modified organisms (GMOs), and earthquake warnings—suggest that it is necessary to rethink how the relationship between science and democracy should be structured. These are not just failures of political practice; they are also failures of political theory.

The tension between expertise and democracy is not a new problem, but it is important to distinguish between two different forms that the problem has taken historically. The first challenges the justification for democratic rule given the alternative of rule by experts. If there are experts who possess superior knowledge about what is best, the argument goes, then having them rule would be in everyone's interest. Participation by those who know less would simply result in worse outcomes for all. This was one of Plato's arguments for philosopher kings, and it is the main claim in recent arguments for epistocracy.[11] The relevant expertise in this case is knowledge of the good or what would be best for the community; it is a form of moral and political knowledge, rather than scientific or technical. This line of reasoning is usually countered by questioning whether such knowledge exists, whether we can identify or agree on

10. In saying this, I align with the "all-subjected principle" on the boundaries of the people. This principle takes bounded political units for granted, but points out that many resident aliens, migrants, and refugees today are unjustly excluded from the political rights and responsibilities of citizenship in states where they are subject to the laws. For a discussion of the all-subjected principle and its more cosmopolitan counterpart, the "all-affected principle," see Sofia Näsström, "The Challenge of the All-Affected Principle," *Political Studies* 59, no. 1 (2011): 116–34. See also Seyla Benhabib, *The Rights of Others: Aliens, Residents, and Citizens* (Cambridge: Cambridge University Press, 2004); Arash Abizadeh, "On the Demos and Its Kin: Nationalism, Democracy, and the Boundary Problem," *American Political Science Review* 106, no. 4 (2012): 867–82.

11. See, for example, Jason Brennan, *Against Democracy* (Princeton, NJ: Princeton University Press, 2016).

those who possess it, whether a small elite or the demos as a whole is more likely to possess it, and if a small elite, whether it can be trusted to rule incorruptibly.[12] These arguments are about the relationship between knowledge and the legitimacy of democratic authority. I mention these only to set them aside.

The other form of the problem of expertise starts from the premise that democracy is a desirable regime type for a variety of reasons and examines the difficulties posed to democratic rule by its inevitable dependence on expertise in policy making.[13] In this case, the expertise in question is scientific or technical. These experts do not claim that they know what is best for the community but rather that they possess the knowledge necessary for attaining democratically determined goals. This book takes up this second form of the problem, and one specific version of it: the relationship between scientific inquiry and politics, where the experts are professional scientists.

The complexity and institutionalization of scientific bodies offering expertise in politics grew rapidly around the middle of the twentieth century.[14] This was a result of the unprecedented alliance between the state and the scientific community following scientists' contributions to the military effort in World War II and the development of the atomic bomb. This new alliance was cemented with the provision of large amounts of public funds for scientific research. The sophisticated, highly professionalized, and expensive scientific enterprise that was established as a result stood in stark contrast with earlier images of science as a largely amateur project. Thinkers such as John Stuart Mill and John Dewey, who were both concerned with the use of scientific expertise in politics, wrote with a different model of scientific inquiry in mind. Science for them was a private activity for curious individuals. The idea of a

12. David Estlund, *Democratic Authority: A Philosophical Framework* (Princeton, NJ: Princeton University Press, 2008); Hélène Landemore, *Democratic Reason: Politics, Collective Intelligence, and the Rule of the Many* (Princeton, NJ: Princeton University Press, 2008); Samuel Bagg, "The Power of the Multitude: Answering Epistemic Challenges to Democracy," *American Political Science Review* 112, no. 4 (November 2018): 891–904.

13. This version of the problem can also be traced back to ancient Greek democracy. For insights on how the ancients dealt with the problem, see Lane, "When the Experts Are Uncertain"; Josiah Ober, *Democracy and Knowledge: Innovation and Learning in Classical Athens* (Princeton, NJ: Princeton University Press, 2008).

14. Michael Oppenheimer, Naomi Oreskes, Dale Jamieson, Keynyn Brysse, Jessica O'Reilly, Matthew Shindell, and Milena Wazeck, *Discerning Experts: The Practices of Scientific Assessment for Environmental Policy* (Chicago: University of Chicago Press, 2019).

professional scientist was a novelty, and many scientists still lacked any kind of formal training.[15]

Scientists today are distinguished by their membership in a professional scientific community. They owe their status and recognition as experts to a complex credentialing system that requires degrees, publications, institutional affiliations, and adherence to professional codes of conduct. Of course, these cannot ensure that scientists will always be experts in an objective sense, or that they will be the right experts to consult for all problems with a scientific dimension.[16] Whether and when scientists are the right experts in a policy context must be determined case by case. Still, the category of the scientist is meaningful as an object of study in politics since scientists are recognized in policy contexts, serve as expert advisers, and have special standing and authority in the public sphere due to their credentials. The existence of a self-regulating and relatively insulated scientific community whose members can have direct influence over the policy-making process thus lends new and distinctive aspects to the old problem of expertise.

The dominant twentieth-century solution to the problem of expertise, developed mostly in the context of social science and especially economics, was to maintain a division of labor between experts and laypeople, modeled after the Weberian account of the relationship between bureaucracy and political leadership.[17] On this view, experts would provide a neutral assessment of the

15. The term "scientist" was coined by William Whewell in 1833. See Laura Snyder, *Reforming Philosophy: A Victorian Debate on Science and Society* (Chicago: University of Chicago Press, 2006).

16. Alvin Goldman argues that an expert is a person who possesses superior knowledge in a given domain than most people, and is able to deploy this knowledge to answer new questions in that domain. Alvin Goldman, "Experts: Which Ones Should You Trust?," *Philosophy and Phenomenological Research* 63, no. 1 (2001): 85–110. This roughly corresponds to Harry Collins and Robert Evans's notion of "contributory expertise." Harry Collins and Robert Evans, *Rethinking Expertise* (Chicago: University of Chicago Press, 2007). These definitions are about what it means to *be* an expert, which may or may not involve social recognition *as* an expert. By contrast, I am interested in those who occupy the social and professional role of scientific expert, and who are recognized as such in the policy context. Of course, the two definitions overlap in many cases; scientists often are the true experts in the areas they study. But they can also come apart. Scientists may be consulted as experts when they are not, and ordinary people may possess expertise according to the Goldman or Collins and Evans criteria, but will usually not be consulted as experts for policy purposes because of their lack of credentials.

17. Max Weber, "Bureaucracy," in *From Max Weber: Essays in Sociology*, ed. Hans Heinrich Gerth and C. Wright Mills (New York: Oxford University Press, 1958), 196–244.

facts, while citizens and their representatives would supply the values necessary for political judgment. Although Max Weber was pessimistic about the ability of bureaucracies to be truly neutral, he held this up as the ideal to strive for. Isaiah Berlin gave a clear expression of this same view in the opening lines of his famous 1958 essay "Two Concepts of Liberty": "Where ends are agreed, the only questions left are those of means, and these are not political but technical, that is to say, capable of being settled by experts or machines, like arguments between engineers or doctors."[18]

Even Jürgen Habermas, who was deeply concerned with the encroachment of scientific and technical expertise into the political sphere, nonetheless accepted the validity of this division of labor. In *Toward a Rational Society*, he deplored the fact that the exigencies of new technologies were increasingly supplanting the decision-making power of political leaders and value judgments were being displaced by the logic of objective necessity.[19] He was concerned that the rationalization of politics would result in science and technology usurping the realm of ends, such that political power would become an empty fiction and all practical matters would be formulated as problems that experts could solve. His solution to this threat of technocracy was to insist on directing scientific knowledge as a means toward goals chosen by deliberating citizens. But he did not question the assumption that experts could be trusted to settle problems about the means in a purely technical and effective way. Although Habermas acknowledged that science is not value free, his conviction in science's capacity for prediction and technological control played a far more important role in his political theory than thorny questions about the epistemic status of scientific claims, which followed from his own pragmatist conception of truth. His one brief mention of uncertainty in this work is revealing. He argued that the reduction of all practical decisions to choice under uncertainty would be the very culmination of rationalization. He failed to note that choice under uncertainty always requires moral judgment—about the outcomes and mistakes that decision makers want to avoid, and the attitudes they take toward risks, which are morally and culturally determined. Scientific probabilities can never determine choice under uncertainty even if we assume that reliable probabilities are available.

18. Isaiah Berlin, "Two Concepts of Liberty," in *Liberty*, ed. Henry Hardy (Oxford: Oxford University Press, 2002), 166–217.

19. Jürgen Habermas, *Toward a Rational Society: Student Protest, Science, and Politics*, trans. Jeremy Shapiro (Cambridge, UK: Polity Press, 1987). See also Jürgen Habermas, *Knowledge and Human Interests*, trans. Jeremy Shapiro (Cambridge, UK: Polity Press, 1987).

These earlier treatments of the problem assumed an idealized view of expertise and were not attentive to the inner workings of science as a practice. They took for granted that experts were successful at providing accurate predictions that enabled rational control over nature. They saw the modern world as characterized by the reduction of contingency; the truly unforeseeable played no part in these theories.[20] Uncertainty was assumed to be probabilistic and subject to human control; nothing, in principle, was beyond scientific prediction. These accounts were driven by the worry that the inexorable logic of technical necessity would crowd out the space for meaningful political choice.[21] Both the Weberian division of labor and Habermas's pragmatic deliberative theory were solutions that aimed to protect a sphere of value-based political judgment beyond the ever-expanding reach of technical assessment.[22]

When we examine recent controversies around scientific knowledge, however—such as on climate change, COVID-19, biotechnology, or artificial intelligence—we see that they are rarely characterized by the predictable decisions, objective assessments, order, rationality, and efficiency that defined twentieth-century hopes and fears around expertise. To the contrary, each case is marked by uncertainty about future outcomes, expert disagreement over the underlying science, and charges of bias on both sides. Many natural processes—like climate change, earthquakes, floods, and hurricanes—are characterized by radical uncertainty, which defies scientific prediction. Further research in these areas often increases uncertainty rather than reducing it and reveals more about what we do not know.[23] The problems of expertise that we

20. Shalini Satkunanandan, "Max Weber and the Ethos of Politics beyond Calculation," *American Political Science Review* 108, no. 4 (2014): 169–81.

21. Sheldon Wolin points out that "the special irony of the modern hero is that he struggles in a World where contingency has been routed by bureaucratized procedures and nothing remains for the hero to contend against. Weber's political leader is rendered superfluous by the very bureaucratic world Weber discovered." Sheldon Wolin, *Politics and Vision: Continuity and Innovation in Western Political Thought* (1960; repr., Princeton, NJ: Princeton University Press, 2004), 379–80.

22. More recent defenders of the division of labor model include Thomas Christiano, *The Rule of the Many: Fundamental Issues in Democratic Theory* (Boulder, CO: Westview Press, 1996); Philip Kitcher, *Science, Truth, and Democracy* (Oxford: Oxford University Press, 2001); Harry Collins and Robert Evans, *Why Democracies Need Science* (Cambridge, UK: Polity Press, 2017).

23. For descriptions of these trends, see Ulrich Beck, *Risk Society: Towards a New Modernity* (London: Sage, 1992); Ulrich Beck and Peter Wehling, "The Politics of Non-Knowing: An Emerging Area of Social and Political Conflict in Reflexive Modernity," in *The Politics of Knowledge*, ed. Fernando Domínguez Rubio and Patrick Baert (London: Routledge, 2012), 33–57;

encounter in these cases do not fit the conceptualization of expertise in twentieth-century accounts.

This book takes the uncertainty, incompleteness, and fallibility of scientific claims to be central to questions about their political use, rather than taking reliable expertise as a black box and asking how it could be used better to advance collective ends. In doing so, I also depart from treatments of the problem of expertise that start from the question of how to improve the public understanding of science in order to use it more effectively. While the public understanding of science is clearly important, starting from this question presupposes that the appropriate role of nonexperts has already been settled, and the primary goal is to inform and educate them about science.[24] If only citizens and policy makers understood the science, the thinking goes, they would be able to make better decisions. This approach puts laypeople in a passive role with respect to the content of expert claims. I propose instead that we start from the prior question of what role citizens and policy makers should play in decisions involving expertise, and will argue that the answer depends on what we know about the limits of expertise. Once we take these into account, we will also arrive at different answers to questions about how science should be translated and communicated, and what form expert-layperson interactions should take.

My central claim in this book is that paying attention to the uncertainty, incompleteness, and possible biases of available scientific expertise as well as the limitations of the decision contexts in which it is used, should change the procedures and institutions appropriate for democratic decision-making on the basis of expertise. Specifically, it gives us reason to make the use of expertise more democratic, flexible, and attentive to the cost and distribution of potential mistakes. This has implications for striking the proper balance between scientific and democratic authority as well as determining the proper procedures for the funding, production, and use of scientific knowledge. The challenge is in specifying what the relevant limitations are, and how and why they should affect democratic procedures. This is the challenge that I take up.

Silvio Funtowicz and Jerome Ravetz, "Science for the Post-Normal Age," *Futures* 25, no. 7 (1993): 739–55; Sheila Jasanoff, *The Fifth Branch: Science Advisers as Policymakers* (Cambridge, MA: Harvard University Press, 1990).

24. Mark B. Brown argues that Kitcher's project is built on the assumption that public distrust and skepticism of science is due to a lack of understanding. See Mark B. Brown, "Philip Kitcher, *Science in a Democratic Society*," *Minerva* 51 (2013): 389–97.

The upshot of the argument is to redraw the boundaries of the Weberian division of labor. While I still assume that professional scientists will be the primary producers of scientific knowledge in society, and nonexperts should supply the goals and value judgments necessary for decision-making, my goal is to expose the fuzzy middle ground where it would be problematic to adhere to a division of labor insofar as facts and values are impossible to separate.[25] The uncertainty and incompleteness of our knowledge and the constraints on decision-making environments mean that certain kinds of judgment that the division of labor model relegates to the expert domain are in fact value-laden ones under uncertainty, and as such, are appropriately made by democratic procedures. Scientists often make these judgments during research or advisory processes, but in doing so they move beyond what is justified by appeal to their superior knowledge. Leaving these judgments unexamined is a failure to exercise a properly democratic responsibility and encourages its inappropriate exercise by experts themselves. I thus argue that nonexperts must scrutinize expert claims, examine the role of values, background assumptions, and uncertainty in the available scientific knowledge, and deliberate about what counts as reliable expertise in a particular context and for a particular purpose. Such scrutiny will be possible only if there are real opportunities for dissent within the scientific community, and institutions that facilitate the discussion and evaluation of science in the public sphere. This requires imagining institutional reforms to reduce obstacles to scientific contestation and prevent the monopolization of knowledge.

A crucial argument of this book is that focusing only on the decision-making stage would give us a narrow picture of how science shapes society and can be used to pursue democratic goals. The alternatives on the table at the decision stage are largely determined by earlier decisions about which research should be pursued and how.[26] These decisions are typically made by funding bodies, which determine not only the direction of scientific and technological change but also the agenda for future political decisions. Once the impact of

---

25. In this I go against recent efforts to draw a boundary between the technical and political, thereby restricting the scope of democratic participation. See Harry Collins and Robert Evans, *Why Democracies Need Science* (Cambridge, UK: Polity Press, 2017); Harry Collins, Martin Weinel, and Robert Evans, "The Politics and Policy of the Third Wave: New Technologies and Society," *Critical Policy Studies* 4, no. 2 (2010): 185–201.

26. Kitcher also draws this connection between the funding and use of science. See Kitcher, *Science, Truth, and Democracy.*

scientific findings becomes clear, nonscientists can accept or reject expert claims, but they cannot procure a different kind of science or wish away existing findings. This book therefore examines structures of funding for science, and considers how to strike a balance between scientific and democratic influence over the distribution of public funds for science. Once again, I start from the uncertainty, incompleteness, and fallibility of our knowledge about the outcome of future research. While this indeterminacy is usually taken to ground arguments for the autonomy of science from political influence, I defend the opposite view and maintain that it can actually lend support to certain forms of democratic intervention in the funding process.

Scholars studying the relationship between science and politics frequently draw a distinction between science for policy and policy for science.[27] The former describes science that informs policy decisions, while the latter focuses on the rules and regulations designed to oversee the conduct of science. This book treats the two as interdependent and traces the implications of the same basic argument in both domains. While the political consequences of scientific findings could perhaps be disregarded by scientists pursuing knowledge in complete isolation from society, they cannot be ignored in a scientific community whose activities are publicly funded and whose findings directly inform policy making.

Two important concerns are worth dispelling from the beginning. The first is that concentrating on the uncertainty and limits of scientific knowledge will devolve into radical skepticism about the ability of science to deliver reliable answers. This will blur the distinction between science and politics, and encourage disregarding expertise and replacing it with common sense. It will become clear in the following chapters that this is not my argument. The starting point of this project is that expert knowledge is indispensable to a modern democracy, and experts have superior knowledge and understanding on many crucial questions of fact. The question of how we should respond to climate change, for instance, cannot be settled by our experience of the weather, nor can it be resolved by deliberating about how much we care about nature or future generations. The answer requires knowing how much the earth will warm, and what the impact will be on different regions. We depend on scientists for these answers.

27. Homer A. Neal, Tobin L. Smith, and Jennifer B. McCormick, *Beyond Sputnik: U.S. Science Policy in the Twenty-First Century* (Ann Arbor: University of Michigan Press, 2008); Heather Douglas, *Science, Policy, and the Value-Free Ideal* (Pittsburgh: University of Pittsburgh Press, 2009).

The point of thinking about the implication of the reliability of scientific claims is not to delegitimize them but instead to be clear about why and how citizens must examine expert claims, and what room there is for democratic judgment on scientific issues. We don't need to believe that science is infallible to make productive use of it. My claim is that what we know about the ways in which it is incomplete and biased should influence the appropriate attitude to take toward knowledge claims, and the correct institutional structures for handling them. Taking heed of the limitations of our epistemic situation does not mean that we should abandon informed decision-making. Economist Robert Solow once remarked that realizing that a perfectly aseptic environment is impossible does not mean one might as well conduct surgery in a sewer.[28] But it also does not mean that we conduct surgery as we would in a perfectly aseptic environment. This project considers how we should change the way we do surgery once we realize that the environment is less aseptic than we believed.

The second concern is that even if this book is careful about the status of scientific claims and the proper balance between scientific evidence and democratic procedures, it might nonetheless have the unintended consequence of increasing mistrust of scientists and disregard for evidence. The argument for democratizing the use of expertise inevitably involves drawing science and scientists onto the political stage and exposing their weaknesses. Given the widespread denial and mistrust of science today, this might embolden those who disregard or discredit scientific evidence. Would it not be more appropriate for theorists today to think of ways to shield expertise from politics rather than opening it up to further scrutiny?

This is a serious challenge, especially since I argue in chapter 6 that researchers bear some responsibility for the unintended but foreseeable consequences of their research. Still, I think it is dangerous to respond to pessimism about the current state of democracy and worries about citizens' ignorance by retreating from democratic principles, and thus removing more and more issues from public input. This response avoids dealing with the root causes of the problem and might lead to a backlash against expertise, as the L'Aquila case demonstrates.

People often feel anxious and fearful about scientific or technological developments because they cannot reconcile new truth claims with their deeply

---

28. Quoted in Clifford Geertz, "Thick Description: Toward an Interpretive Theory of Culture," in *Readings in the Philosophy of Social Science*, ed. Michael Martin and Lee McIntyre (Cambridge, MA: MIT Press, 1994), 230.

held values and cultural commitments. Scientific claims do not intrinsically favor one worldview or set of values over another, but scientists and others who produce and translate scientific findings for use in public life wield significant power in determining which worldviews or values will appear compatible with scientific knowledge. If decisions about which findings are accepted as true for political purposes and what knowledge becomes available for use are removed from democratic influence, citizens might find themselves reduced to a choice between deference to the judgments of scientists and a rejection of the authority of science altogether. This disempowers the public, and encourages unaccountable and irresponsible policy making. Expanding the possibilities for democratic engagement over science is a way to avoid this stark choice and open up more flexible options for reconciling science with politics. This, in turn, can only be done by reinvigorating existing democratic institutions and imagining new ones.

A View of the Theoretical Landscape

The relationship between science and democracy has been examined mostly by scholars in science and technology studies (STS) and the sociology of science along with a few philosophers of science. As a political theorist, what distinguishes my approach is that I place political institutions at the center of my analysis. Scholars in STS have usually avoided thinking in terms of institutions and been particularly wary of taking a normative stance.[29] Similarly, most philosophers of science who have demonstrated how social values shape scientific findings have stopped short of tracing the systemic political implications of these important results. I follow the example of philosophers Philip Kitcher and Heather Douglas, whose pioneering work bridges the gap between the philosophy of science and political philosophy. But they too have largely neglected the dynamics of existing political structures at the intersection of science and policy. While Kitcher develops a highly idealized model

---

29. See, for example, Jasanoff, *The Fifth Branch;* Sheila Jasanoff, ed., *States of Knowledge: The Co-Production of Science and Social Order* (London: Routledge, 2004); Alan Irwin and Brian Wynne, *Misunderstanding Science? The Public Reconstruction of Science and Technology* (Cambridge: Cambridge University Press, 1996); Bruno Latour, *Politics of Nature: How to Bring the Sciences into Democracy* (Cambridge, MA: Harvard University Press, 2004). For a similar critique of the STS literature for failing to engage with political theory, see Alfred Moore, "Beyond Participation: Opening Up Political Theory in STS," *Social Studies of Science* 40, no. 5 (2010): 793–99.

based on hypothetical discussions between scientists and "tutored" citizens, Douglas examines political problems around science and democracy through the lens of the individual moral responsibilities of scientists.[30]

The distinctive contribution of this book is to develop a theory through the close examination of three types of formal institutions that play a crucial role in the production and use of scientific expertise in democratic societies: scientific advisory committees, small-scale democratic experiments, and funding bodies. I explore their internal dynamics and broader democratic role through the framework of conceptual concerns around authority, legitimacy, equality, freedom, representation, accountability, and inclusion, which are part of the standard vocabulary of political theory. While the thrust of my argument is broadly consistent with works in STS and philosophy that have argued for the need to democratize science, I try to be more precise about what this general claim means for specific institutional bodies and the actors within them. To this end, I articulate specific dilemmas that arise at the intersection of science and democracy—between scientific neutrality and political usefulness, expert knowledge and public participation, scientific autonomy and democratic control, and freedom of inquiry and protection from harm—and reflect on how to resolve them.

The design of institutions requires empirical evidence about performance and information about the particulars of a context, which go beyond the scope of a largely theoretic project such as this one.[31] The same institutions will not be appropriate for all democracies at all times, and we cannot predict the performance of institutions purely from their design. The institutional suggestions I make in this book should therefore not be interpreted as all-things-told prescriptions meant to apply regardless of time and place but rather as practical illustrations of the theory. The aim is to demonstrate that my arguments about the proper relationship between science and democracy *could* be institutionalized, and spell out which institutional forms would better realize them and why. Taking up the challenges of practical specification strengthens a

30. Kitcher, *Science, Truth, and Democracy*; Douglas, *Science, Policy, and the Value-Free Ideal*; Heather Douglas, "The Moral Responsibilities of Scientists (Tensions between Autonomy and Responsibility)," *American Philosophical Quarterly* 40, no. 1 (January 2003): 59–68.

31. For empirical studies of these institutions from a political science / science policy perspective, see David Guston, *Between Politics and Science: Assuring the Integrity and Productivity of Research* (Cambridge: Cambridge University Press, 2000); Roger Pielke Jr., *The Honest Broker: Making Sense of Science in Policy and Politics* (Cambridge: Cambridge University Press, 2007).

theory, even if a gap between theory and practice always remains. In the end, a theory must make some assumptions and idealizations, which place conditions on its applicability to particular contexts. This is a point that I emphasize about the use of scientific models in policy; it is only fair to acknowledge that it also applies to my own theory. I try to respond to this by explaining my assumptions, and supporting their plausibility with relevant evidence and examples where possible.

While other fields have studied the relationship between scientific expertise and democracy using their own theoretical and methodological frameworks, political theorists have largely neglected the subject. My book joins two excellent ones published by political theorists Mark Brown and Alfred Moore in recent years in an effort to establish the problems of scientific expertise as a vital area of inquiry for contemporary democratic theory.[32] Although I agree with many of their points, this book departs from theirs in emphasizing the role of decisions made during the funding and research stages in shaping the political agenda and constraining the possibilities for action. I therefore devote attention to earlier stages of the research process, and argue that the democratization of expertise must be rooted in the democratization of decisions about what kinds of knowledge are pursued and how. My concern with the funding and design choices made at the research stage is absent from these books. I should add that I reject the label of elitism, which Moore uses to describe his theory; to the contrary, I maintain that democracies must find ways to reclaim some of the elite power over agenda setting and decision-making for ordinary citizens themselves.

My position could be described as aiming to democratize the use of expertise, but it is important to clarify how it differs from recent arguments in favor of epistemic democracy and the wisdom of crowds.[33] Unlike scholars in this area, I make no claim that democratic procedures are on the whole more likely to produce "correct" outcomes than decision-making by experts in domains involving complex expert knowledge. The undeniable asymmetries in

32. Mark B. Brown, *Science in Democracy: Expertise, Institutions, and Representation* (Cambridge, MA: MIT Press, 2009); Alfred Moore, *Critical Elitism: Deliberation, Democracy, and the Problem of Expertise* (Cambridge: Cambridge University Press, 2017).

33. Landemore, *Democratic Reason*; Robert Goodin and Kai Spiekermann, *An Epistemic Theory of Democracy* (Oxford: Oxford University Press, 2018); James Bohman, "Deliberative Democracy and the Epistemic Benefits of Diversity," *Episteme* 3, no. 3 (2006): 175–91; Elizabeth Anderson, "The Epistemology of Democracy," *Episteme* 3, nos. 1–2 (2006): 8–22.

the knowledge possessed by a small group of experts and ordinary citizens make epistemic arguments for democracy difficult to carry over to domains of complex expert knowledge.[34] More important, my argument suggests that on issues involving expertise, assessments of the quality of decisions will be determined by the certainty, completeness, and bias of the available expert knowledge. The scientific questions that have been pursued—and how they have been pursued—place limits on the kinds of decisions that citizens will consider possible or desirable on issues where the need for scientific knowledge is acknowledged.

A growing literature on bureaucracies and the administrative state within political theory addresses some problems of expertise.[35] I share this literature's goal of directing theorists' attention to the inner workings of democratic government and administration. Despite some basic similarities, however, the role of science in politics cannot be explained fully by theories designed to analyze the role of bureaucracies. The fundamental worry about bureaucratic domination does not apply well to scientists since they are rarely delegated power to make binding rules. Scientists do not occupy political positions that allow them to exercise arbitrary power over other citizens.[36] The authority they possess is usually epistemic, advisory, or cultural. This does not mean that their authority is unproblematic, but articulating when and why scientific authority becomes a source of democratic concern is a distinct theoretical challenge, which cannot be subsumed under the problem of bureaucratic domination.[37]

---

34. Landemore defines the scope of her argument as "political decisions," but admits that it is not likely to apply to complex problems such as climate change. See Landemore, *Democratic Reason*.

35. Henry S. Richardson, *Democratic Autonomy: Public Reasoning about the Ends of Policy* (New York: Oxford University Press, 2002); Sabeel Rahman, *Democracy against Domination* (New York: Oxford University Press, 2016); Pierre Rosanvallon, *Democratic Legitimacy: Impartiality, Reflexivity, Proximity*, trans. Arthur Goldhammer (Princeton, NJ: Princeton University Press, 2011); Bernardo Zacka, *The State Meets the Street* (Cambridge, MA: Harvard University Press, 2017); Chiara Cordelli, *The Privatized State* (Princeton, NJ: Princeton University Press, 2020); Leah M. Downey, "Delegation in Democracy: A Temporal Analysis," *Journal of Political Philosophy* (2020), doi.org/10.1111/jopp.12234.

36. Scientists may have arbitrary power over human subjects of research and physicians over their patients unless these relationships are well regulated. I bracket these to focus on the role of scientists in politics.

37. For a defense of the view that scientists do not pose a problem for liberal democracy, see Stephen Turner, "What Is the Problem with Experts?," *Social Studies of Science* 31, no. 1 (2001): 123–49.

The principal-agent framework commonly used to analyze the relationship between bureaucracies and legislatures does not apply straightforwardly to scientists either. Scientists are not the agents of politicians or the public, except when they take up certain advisory offices. Even then, they remain highly independent actors constrained mainly by professional incentives and norms. Their proper role with respect to democratic aims and the extent of their answerability to the public must be theorized rather than assumed. In fact, if the default assumption about bureaucracies is that they ought to be subject to legislative control, the default assumption about scientists is that they ought to be free from democratic control. Of course, this simple contrast misses the complex interdependence between science and democracy that this book examines, but it shows how much variation there is among the democratic expectations from actors loosely categorized as experts.

Scope of the Argument

A few clarifications about the scope of the argument are in order. This project focuses on the natural sciences and largely brackets the social sciences. The distinction is admittedly arbitrary since the philosophical views of science that I draw on challenge the conventional distinction between natural and social sciences as value free and value laden, respectively. It is more accurate to treat the natural and social sciences as continuous rather than different in kind. Still, there are two mainly practical reasons for drawing this line. First, this distinction is commonly made both in theory and practice. Philosophers of science typically concentrate on one or the other, or compare the two with the assumption that they are distinct enough in subject matter and the methodological challenges they face. Political institutions such as legislative committees, executive agencies, advisory bodies, and funding institutions also treat these two areas separately.

Second, even if the natural and social sciences lie on a spectrum, the social sciences lie at the end of the spectrum where predictions are less reliable, well-established findings are fewer, and concept formation and measurement are more difficult. There are well-known methodological challenges specific to explaining and predicting human behavior. On the one hand, these factors might make the social sciences a more fruitful, less controversial, and overall easier target for a book that starts from an epistemological critique to argue for the democratic scrutiny of science. On the other hand, the same reasons make the social sciences a less challenging and rewarding subject for study because I suspect that few would disagree with the conclusions. If my argument succeeds in the case of the natural sciences, then a fortiori, it applies to the social sciences too.

Another clarification concerns the applicability of the argument within the natural sciences. Is it meant to apply to all natural sciences or only to some? Do we want democratic participation on all issues or can we leave some safely to experts? These questions are more difficult to answer in the abstract because they depend importantly on which scientific issues become politicized and how. The easy part of the answer is that the argument applies to science that has some relevance to policy. It is not concerned with science in the lab that acquires no relevance for public affairs, except for the discussion of funding for basic research in chapter 5. Within areas of science that acquire policy relevance, I think the argument will be most salient on issues that are highly uncertain, with many unknowns and inadequate evidence, and where the political stakes are high. Although we could try to classify sciences according to their level of certainty—with earthquake and climate science, for instance, being less certain than physics or chemistry—it would be a mistake to try to be specific about which particular scientific areas are likely to fall in this category. I do not mean to suggest that every technical issue should be politicized—if a bridge needs to be built, we could safely leave it to engineers—but rather that the question of which issues should or will be politicized is not one that can be specified in theory.[38]

The point about uncertain and high-stakes science suggests another reason why this project is timely: the big scientific problems of our time—COVID-19 and climate change—have been marked from the beginning by a high degree of uncertainty and disagreement among scientists as well as high political stakes. That there is anthropogenic climate change may not be in dispute among scientists anymore, but the key policy-relevant details about how much warming there will be and how it will affect different regions remain unclear. Different climate models prioritize different epistemic values, and make different background assumptions about the historical record, future human behavior,

---

38. Even bridges can be controversial. The collapse of a bridge in Genoa set off a bitter controversy: Was the accident due to the fallibility of engineering science, or had the management company and Ministry of Infrastructure been negligent? See James Glanz, Gaia Pianigiani, Jeremy White, and Karthik Patanjali, "Genoa Bridge Collapse: The Road to Tragedy," *New York Times*, September 6, 2018, https://www.nytimes.com/interactive/2018/09/06/world/europe/genoa-italy-bridge.html.

and the relative importance of different risks. These features make it clear why democratic engagement must be partly over the content of the science and involve some scrutiny of competing models rather than a debate about moral values that could be addressed independently from the facts. This has not always been the character of the scientific issues that have commanded political attention. The most important scientific issues on the political agenda after the Second World War—the bomb and the space program—were cases where the science was not in dispute. The dilemmas they raised were moral ones about the responsible use of the science. If the division of labor model seemed appropriate for the scientific problems of those times, the more thoroughly democratic model proposed in this project will be more appropriate for ours.

I should add that this book does not focus on the strategic distortion and manipulation of scientific research by corporations, or fabrication of results by individual scientists. These are serious and widespread problems, but they have been documented and analyzed by other scholars.[39] Although more work must be done to reduce their prevalence, I think this work is primarily practical, not philosophical. The problems that I will explore are the ones that remain even when scientists advise policy makers in good faith and intend to solve problems that depend on scientific knowledge. I think this is a realistic description of many expert advisory committees composed of independent research scientists, which are impeded less by deception and fraud, and more by the limits and uncertainty of scientific knowledge and disagreements over values.

Having clarified which kinds of science fall within the scope of this book, I should also say something about how I define democracy. The arguments in this book are intended to be compatible with most widely held normative views of democracy rather than aligning with one specific conception. I take democracy to be a regime characterized by political equality, where collective decision-making procedures are arranged so as to give everyone subject to decisions an equal right to participate in their making.[40] In modern

---

39. This literature is vast. Highlights include Naomi Oreskes and Eric Conway, *Merchants of Doubt: How a Handful of Scientists Obscured the Truth on Issues from Tobacco Smoke to Global Warming* (New York: Bloomsbury Press, 2010); Robert Proctor and Londa Schiebinger, eds., *Agnotology: The Making and Unmaking of Ignorance* (Stanford, CA: Stanford University Press, 2008); Thomas McGarity and Wendy Wagner, *Bending Science: How Special Interests Corrupt Public Health Research* (Cambridge, MA: Harvard University Press, 2008).

40. For similar definitions, see Robert A. Dahl, *Democracy and Its Critics* (New Haven, CT: Yale University Press, 1989); Christiano, *The Rule of the Many;* Niko Kolodny, "Rule over None I: What Justifies Democracy?," *Philosophy and Public Affairs* 42 (2014): 287–336.

representative democracies, this principle is typically institutionalized through free, fair, and contested elections, universal suffrage, guarantees for basic political freedoms of speech, assembly, and association, and popular control over elected representatives.[41] This definition should make my arguments compatible with many liberal, republican, deliberative, participatory, and radical accounts, if not all, but not with elitist theories that view public participation between elections as unnecessary or undesirable. It may also leave out some purely instrumental accounts that take outcome-based criteria to be constitutive of democracy. While my definition of democracy is not an instrumental one, starting from the assumption that modern democracies depend on expertise is to assume an instrumental interest in bringing about good outcomes. The theoretical puzzles in this book do not even get off the ground without this assumption.

It is also worth clarifying the distinction between the role of science in a democracy versus in politics. This book assumes the overall preferability of democracy to other regime types without defending it anew and focuses on the question of how democracies can handle expertise better. Insofar as authoritarian regimes depend on expertise too, some of the answers I provide about the proper division of labor between politicians and experts could be exported to nondemocracies. An example from chapter 3 takes up a gridlock between Napoléon and his main corps of engineers that resulted from a prolonged scientific dispute over the construction of a new canal. The book's main argument, though, has a normative core that does not carry over comfortably to an authoritarian context. It ultimately rests on a view about whose judgments and values should shape the policies by which a community lives. The argument is concerned with the proper source of authority over decisions of a certain kind and the relationship between scientific and democratic authority in decisions that depend on both. These are important in a democratic regime because democracies are meant to be responsive to the values and preferences of their citizens, even if they fall short of this ideal in practice. In authoritarian regimes that make no such claim even in theory, the problem of expertise can be reduced to questions about efficiency and effectiveness. The concern is over how the state can use expertise effectively, and whether decision-making by experts or career politicians produces "better" outcomes. The question of

41. The precise meaning of popular control is controversial. For one answer that I find persuasive, see Sean Ingham, *Rule by Multiple Majorities: A New Theory of Popular Control* (Cambridge: Cambridge University Press, 2019).

which kind of unelected person should make decisions for the public does not have quite the same normative edge; the more pressing question is why those who must obey the state's decisions are given no say at all.

Moreover, the main tensions animating the relationship between democracy and science are due to the existence and social authority of an independent, autonomous scientific community, which is free to pursue knowledge in areas of its own choosing and share the result of its inquiries publicly. Hypothetically, an authoritarian government committed to protecting the autonomy of the scientific community might encounter similar tensions between science and politics. In reality, authoritarian regimes too often lapse into directing, controlling, or repressing the activities of scientists, or worse, silencing, imprisoning, or exiling scientists themselves. The fact that this possibility is widely known in turn determines what scientists dare to do in such contexts, even when the state does not interfere. The interesting tensions in the relationship between science and democracy have essentially been dissolved in nondemocracies in favor of the dominance of politics. The role of science in authoritarianism raises different and interesting tensions of its own. These, however, lie beyond the scope of this book.

Plan of the Book

The book is organized as follows. Chapter 2 develops a taxonomy of the different ways in which the values and purposes of scientists influence their findings, and demonstrates how this affects the practical use of findings later on. Drawing on recent work in the philosophy of science as well as case studies of climate modeling, AIDS, GMOs, medical research, acid rain, and COVID-19, I focus on choices about the formation of concepts, development of hypotheses, construction of models, selection of evidence, and design of experiments. At each stage, scientists make judgments or assumptions about what is significant, useful, or relevant knowledge, and weigh the acceptability of different kinds of mistakes under uncertainty with specific scientific or practical purposes in mind. These judgments favor some perspectives and purposes over others by determining what is known and how it is known. Failure to detect and respond to the way in which these assumptions and values shape expertise will result in democratic policies being influenced imperceptibly by unexamined scientific choices. To address this problem, I argue that democratic institutions that rely on expertise should be oriented toward exposing the assumptions and values driving expert claims, pay special attention to the gaps in the existing body of knowledge, and be more deliberate about the kinds of new knowledge that should be pursued and used.

Chapter 3 considers the translation of science for use, and the relationship between scientific and political authority by analyzing the role of scientific advisory committees in politics. What I call the paradox of scientific advice consists in the fact that the expectation that scientific committees must be politically neutral is not fully compatible with their fundamental task of providing useful advice to inform policy. To help decision makers set and attain democratic goals, scientific advice must be relevant, compatible with citizens' values and purposes, responsive to preferences over risks and errors, and simplified in ways that facilitate rather than preempt democratic judgment. Converting good science to good advice requires making assumptions about ends and values, and thus violates the neutrality that is the source of the authority of scientific bodies. Scientific committees can respond to this dilemma either by sticking to technicalities and risking irrelevance or making value judgments in the name of other citizens, thereby raising concerns about the inadequacy of their claims to representation. I discuss the shortcomings of both these alternatives and suggest that these tensions could be mitigated by strengthening democratic scrutiny through modes of organized scientific dissent directed toward a public audience.

Chapter 4 develops an institutional proposal to facilitate the kind of democratic scrutiny over expertise argued for in chapters 2 and 3. I highlight three main challenges to democratic debate on complex scientific issues: ordinary citizens cannot set the agenda and terms of the debate, they face difficulties evaluating competing expert claims because of their lack of expertise, and asymmetries in knowledge and authority make deliberation between experts and laypeople unproductive. To address these three challenges, I develop a proposal for an adversarial "science court" that would be initiated by citizens, and where experts would be invited to make the case for different views on a scientific question. A citizen jury would question the experts, and then deliberate and deliver a decision. The outcome of the court would serve an advisory role in the policy-making process and inform public debates. The adversary structure of the proposal is designed to expose the background assumptions, potential biases, and omissions in rival expert claims as well as to clarify the levels of uncertainty. The separation of scientist advocates from citizen jurors avoids the difficulties of mutual deliberation under conditions of unequal authority, while allowing citizens to be active participants despite their lack of expertise. The chapter ends by discussing the court's democratic status and legitimacy, and responding to objections about citizens' competence.

The possible uses of science at the decision stage are shaped by earlier decisions about which research areas should be pursued. Chapters 5 and 6 thus turn to institutions for the funding of scientific research as potential sites of longer-term and more foundational democratic input into the political role of scientific expertise. They focus on how funding decisions for science should be made, and what kinds of political interventions would be justifiable and desirable at these earlier stages.

Chapter 5 asks whether there should be democratic input into decisions about the distribution of funds among scientific projects, and if so, what kind and on what grounds. I develop the argument through an examination of two justifications offered for publicly funded scientific research after World War II. The first is engineer and science administrator Vannevar Bush's vision of the universal material benefits from scientists pursuing basic research. Bush followed scientist-turned-philosopher Michael Polanyi in claiming that these benefits would be best realized if scientists were given a high degree of autonomy to pursue their curiosity. I then turn to John Rawls's more modest justification of public funding for science on the benefit principle, which was intended to ensure that individuals paid only for the benefits they wanted. Despite their differences, both accounts failed to consider the political impact and uses of scientific research. I argue that the close connection between scientific inquiry and truth, and special link between science and policy in the modern state, provide additional reasons for the public funding of science that go beyond those that apply to ordinary public goods such as roads and bridges. I sketch an alternative justification for funding science, rooted in the shared democratic interests of citizens in bringing about good outcomes, setting the political agenda, and acquiring the knowledge and competence to hold policy makers accountable on technical issues.

Chapter 6 ventures into more controversial territory, and asks whether and when democracies may restrict or ban certain kinds of scientific inquiry altogether. My goal is to offer a framework for deciding whether to restrict research under conditions of empirical and normative uncertainty, focusing in particular on the appropriate interaction between expert-led and democratic processes. The argument starts from the ethical framework for the regulation of scientific research with human subjects. It is widely accepted today that research may be restricted if it poses harm to human subjects participating in the research process. Far more controversial is the suggestion that research may be restricted on the grounds that the findings pose a risk of harm to society, even if the research is ethically conducted and the findings are true. I argue that this boundary is arbitrary from a moral perspective, and consider how the framework's key principles of beneficence and respect may be adapted for the purposes of considering the broader category of harms to society from the use of scientific knowledge and its application in technology. To do so, I defend a more robust understanding of responsibility that is sensitive to the context in which scientific research takes place and involves assigning scientists some responsibility for the foreseeable consequences of their research, even if they themselves neither inflict nor intend harm. I also maintain that a democratic society would be justified in preemptively restricting research on the basis of collective fear and anxiety under conditions of indeterminacy.

Chapter 7 traces the implications of the argument for the public trust in science, science communication, and the role of scientists in public life, and offers concluding reflections. Finally, chapter 8, an epilogue on the COVID-19 pandemic, shows how the questions addressed in each chapter of the book—from the role of values and uncertainty in science to the paradoxes of scientific advice, from the need for public participation to the role of democratic input into decisions to fund science—became salient in the COVID-19 context. It aims to demonstrate the critical and clarificatory power of my arguments for making sense of this episode, while providing concrete illustrations of ideas discussed more abstractly in other chapters of the book. The complex scientific and political dynamics of the COVID-19 pandemic, in turn, allow me to refine the details of my arguments, and add a few nuances and caveats.

2

Significant Knowledge

IT IS COMMON for medical researchers today to study biological differences across social groups. The effects of diseases, drugs, and vaccines are explained with respect to sex, race, ethnicity, or age differences. During the COVID-19 pandemic, for example, the observation that ethnic minorities were more likely to be admitted to critical care units and exhibited higher rates of death on hospitalization immediately led to calls for more research into the sources of biological differences across ethnic groups.[1] Funders may even require research proposals to include plans for studying impacts across groups. The focus on biological differences across social groups influences the questions that researchers ask as well as the methods and approaches that they use to answer them.

But this paradigm was not always dominant in medicine. Until a few decades ago, most researchers considered humans to be biologically the same for medical purposes. This was a background assumption in the discipline rather than a scientifically verified or philosophically defended claim, and yet it determined shared beliefs among researchers about what constituted good medical research. Only the racists and sexists included variables of social group differences in their studies, and this research was typically regarded as pseudoscientific and driven by a desire to establish the innate inferiority of some humans. Sociologist Steven Epstein demonstrates that it was members of marginalized groups themselves who pushed for the inclusion of social group variables in mainstream medical research over the years.[2] The eventual change

---

1. Manish Pareek, Mansoor N. Bangash, Nilesh Pareek, Daniel Pan, Shirley Sze, Jatinder S. Minhas, Wasim Hanif, and Kamlesh Khunti, "Ethnicity and COVID-19: An Urgent Public Health Research Priority," *Lancet* 395, no. 10234 (May 2020): 1421–22.

2. Steven Epstein, *Inclusion: The Politics of Difference in Medical Research* (Chicago: University of Chicago Press, 2008).

in paradigm from sameness to what Epstein calls inclusion and difference did not result from changes in scientific knowledge and understanding but rather from the external pressures of a social movement that insisted on the relevance of these variables to good scientific explanation.

The inclusion of social variables led to predictable improvements in the medical understanding of the effects of diseases and drugs on women and minority groups, which in turn led to some improvements in health outcomes for these groups. Epstein, however, does not present this purely as a narrative of progress. He argues instead that the inclusion and difference paradigm is also a partial one, much like the earlier paradigm of sameness.[3] There is no scientific justification for why these social identity variables and not others should be the most salient for understanding biological differences across humans. It remains possible that other variables currently not regarded as salient by the medical community could play a more important explanatory role in the future. The hypotheses considered plausible depend on what the medical community considers significant, which is shaped against a set of values and social beliefs that operate in the background of research.

The body of existing scientific knowledge is but a small part of the vast amount of knowledge we could possess. Our epistemic situation is defined as much by what we don't know as what we know. Completeness in knowledge is an impossible ideal. But scientific inquiry is not merely a random accumulation of truths either. As philosophers of science have pointed out, science aims to seek out and organize a body of *significant* knowledge.[4] What constitutes significance is determined with reference to a value scheme assumed in the background of inquiry. It may be derived from the constitutive purposes of professional scientific disciplines, specific practical or technological goals, or the moral, cultural, and political values of individual scientists. These judgments of significance may not affect the truth or empirical reliability of scientific claims, but they determine the contours of the body of scientific knowledge in a society at a given time. They influence not only the selection of questions but also the definition of concepts, choice of methodologies, construction of theories, design of experiments, and selection of evidence.

---

3. Epstein, *Inclusion*.

4. Elizabeth Anderson, "Knowledge, Human Interests, and Objectivity in Feminist Epistemology," *Philosophical Topics* 23, no. 2 (Fall 1995): 37; Philip Kitcher, *Science, Truth, and Democracy* (Oxford: Oxford University Press, 2001), 63–85.

The aim of this chapter is to describe and illustrate the role of significance judgments at different stages of scientific inquiry, and articulate why and how they matter for the relationship between science and democracy. The first half of the chapter develops a taxonomy of the different ways in which the values and purposes of scientists shape their methods, theories, and findings, and shows how these values and purposes affect the practical use of findings later on. Scientists make judgments or assumptions about what is significant, useful, or relevant knowledge, and weigh the acceptability of different kinds of mistakes with certain scientific or practical purposes in mind. These judgments favor some perspectives and purposes over others in determining what is known and how it is known.

The second half of the chapter discusses the implications of this for how democratic societies use scientific expertise in decision-making. My main claim is that the way scientists study an issue determines how it is understood in a society, frames public debate, and influences the policies that appear feasible or even conceivable. What scientists deem insignificant or overlook altogether constitutes the gaps in a society's knowledge, while scientists' biases become the biases of decisions taken on the basis of scientific knowledge.[5] The problem is that this influence of science over politics operates through what is taken as background technical knowledge rather than as part of an open democratic deliberation over ends. To prevent democratic decisions from being shaped by experts' unexamined and necessarily partial perspectives, I argue that institutions and decision procedures that rely on scientific expertise should be designed to expose and respond to these values, and be more attentive to the limits, biases, and incompleteness of science. This requires submitting scientific expertise to critical democratic scrutiny to understand how scientific assumptions and purposes shape findings, and test their adequacy for the attainment of democratic purposes.

A Taxonomy of Values in Science

It is widely accepted that science is shaped by social values and purposes. What is more contentious is how values shape science and what follows from this fact. I do not intend to contribute directly to the philosophical debate about the proper role of values in science but instead to survey this literature

---

5. Robert Proctor and Londa Schiebinger, eds., *Agnotology: The Making and Unmaking of Ignorance* (Stanford, CA: Stanford University Press, 2008).

with the aim of identifying the arguments that have the most important implications for the relationship between science and democracy. My main aim is to articulate and extend the political implications of these philosophical arguments.

It is helpful to start this task with a quick taxonomy of the role that values can play at different stages of scientific inquiry. We can divide up the process of scientific inquiry schematically into five stages: the selection of research questions, selection of methodology and research design, collection of data, acceptance or rejection of a hypothesis, and application of findings to solve a practical problem. The selection of questions and application of findings are sometimes called the external stages of inquiry, while the selection of methodology, research design, data collection, and acceptance or rejection of hypotheses are regarded as the internal stages. Following convention, I will treat the internal and external stages separately, focusing only on the internal stages in this chapter and leaving the external stages to later chapters. Chapters 3 and 4 take up the application of findings to practical decisions, while chapters 5 and 6 focus on the selection and funding of research questions.

The most heated debates among philosophers revolve around the question of whether it is acceptable for values to play a role in the judgment that the evidence supports a hypothesis. This is controversial because a direct role for values at this stage would jeopardize the empirical reliability of science.[6] The fact that moral or political values favor a hypothesis is not evidence for its truth, and does not provide the right kind of reason to accept it. I will simply grant that it is illegitimate to give values a direct role at this stage.[7] Even if the judgments of individual scientists are necessarily influenced by their own values and perspectives, the mechanisms of collective criticism and scrutiny within the scientific community can—and should—be oriented toward detecting and challenging any objectionable role that values play in the acceptance of scientific theories.[8] The proper functioning of these internal mechanisms of scrutiny can in principle ensure the reliability of scientific claims.

6. Heather Douglas, *Science, Policy, and the Value-Free Ideal* (Pittsburgh: University of Pittsburgh Press, 2009), 96.

7. I follow Douglas's use of the word "direct" to mean providing warrant for acceptance in itself rather than supporting acceptance indirectly through the judgment that the evidence is sufficient.

8. Helen Longino, *Science as Social Knowledge: Values and Objectivity in Scientific Inquiry* (Princeton, NJ: Princeton University Press, 1990), 62–82.

Instead of focusing on the acceptance or rejection of hypotheses, I want to shift attention to a range of value judgments made earlier during the research process. Assumptions about what counts as significant knowledge and how research should be conducted to produce significant knowledge are usually made at earlier stages. They most obviously affect the selection of research questions, but also shape the selection of methodologies, definition of concepts and variables, construction of theories and models, and standards about what kinds of evidence are acceptable and relevant. These judgments are not usually considered problematic from a scientific perspective because they do not threaten the relationship between evidence and hypothesis, which is at the core of scientific inference. As a result, far less philosophical attention is devoted to examining how these earlier judgments shape the existing body of scientific knowledge, and what this implies for their social impact and use. I follow Kitcher in taking these judgments about significance to have the most crucial implications for the relationship between science and democracy, but I include a broader range of methodological and theoretical choices in this category, going beyond his focus on the selection of questions. In doing so, I show that the influence of these judgments over the body of existing scientific knowledge is both more far reaching and difficult to detect.

Kitcher builds on earlier work by philosophers such as Nancy Cartwright and John Dupré to argue against an influential view that maintains that the significance of scientific truths can be derived from the intrinsic structure of the world—for instance, through their hierarchical relationship to the fundamental laws of nature.[9] In place of this view, he offers an alternative where scientific significance should be understood as derived from "our contingent and evolving interests."[10] To ensure that scientific knowledge can further democratic ends, he goes on to propose an ideal in which the scientific community would pursue truths whose significance would coincide with judgments that would be reached in an ideal deliberation among representatives of all viewpoints in society. In pitching his theory at a high level of idealization, however, Kitcher fails to engage with the political implications of the way scientific questions are asked and answered in currently existing societies. His utopian model avoids the difficult political questions for here and now about

9. Nancy Cartwright, *How the Laws of Physics Lie* (New York: Oxford University Press, 1983); John Dupré, *The Disorder of Things: Metaphysical Foundations of the Disunity of Science* (Cambridge, MA: Harvard University Press, 1995).

10. Kitcher, *Science, Truth, and Democracy*, 72.

how value judgments made during actual research processes favor some perspectives over others, and what this means for real democracies rather than hypothetical "well-ordered" ones.

My aim is to revisit the role of value judgments throughout the research process to expose how they facilitate the pursuit of some ends and satisfaction of some needs over others. I then intend to diagnose the democratic problem posed by our reliance on a body of knowledge shaped by these judgments. In practice, judgments of significance are determined by the constitutive purposes of specific disciplines, their shared norms, interests, and background assumptions as well as by scientists' own moral and political concerns. They may be derived from specific practical purposes, such as protecting ecosystem balance or exerting technical control over nature, or reflect professional purposes, such as expanding a particular research program. From a democratic perspective, it is crucial to understand how different practical pressures influence the design and conduct of scientific inquiry because these choices influence the political decisions that are possible to pursue on the basis of the existing body of scientific knowledge and solutions to social problems that are even conceivable. Significant omissions, in turn, determine the gaps in a society's body of knowledge.[11] These effects are not subject to correction by the self-correcting mechanisms of science because there may be nothing to correct in a judgment of significance from a scientific perspective.

In the following sections, I will illustrate how values shape scientific research at early stages through a discussion organized thematically around concepts, hypotheses, models, evidence, and randomized control trials.

Concepts

One of the earliest stages at which scientific inquiry can incorporate values is in the definition of concepts. While some concepts are purely factual and others purely evaluative, many widely used concepts combine empirical observation with normative judgment.[12] Descriptions of environmental or biological changes as "damage" or "harm," and the characterization of certain

---

11. Proctor and Schiebinger, eds., *Agnotology.*

12. Hilary Putnam, *The Collapse of the Fact/Value Dichotomy* (Cambridge, MA: Harvard University Press, 2002); John Dupré, "Fact and Value," in *Value-Free Science: Ideals and Illusions?*, ed. Harold Kincaid, John Dupré, and Alison Wylie (New York: Oxford University Press, 2007), 27–42.

possibilities as "risks" or "dangers," are some examples.[13] These terms can be viewed as "thick" concepts, following Bernard Williams's usage of the phrase to describe words that contain both factual and ethical content, such as cruel, courageous, and elegant.[14] In the scientific context, thick concepts combine an observational quality, which is derived from scientific research, with an evaluative quality, which classifies observations as good or bad according to an implicit normative framework that defines their significance. The judgment that certain environmental or biological changes constitute damage, harm, danger, vulnerability, disease, or risk requires assuming a baseline for what is normal for that system as well as an evaluative judgment about which departures from this baseline are good or bad. The latter judgment cannot be made neutrally, or without reference to a subject or purpose. Good for whom or what? The answer will depend on whether the perspective of humans, animals, plants and ecosystems, local residents, industry, or neighboring countries is adopted, and how each perspective is represented.

The use of a thick concept to describe a natural phenomenon entails a commitment to the underlying normative framework and its practical implications. To claim that a disease has spread in a forest is not only to describe a change in a natural system and signal that this change is bad but also to imply that commonly held normative views about how to respond to disease are appropriately invoked in this context. For instance, describing a forest as diseased might imply that restoring it to a state of health would be desirable if the costs are not onerous. Similarly, to describe a tumor as malign is to signal that it is bad for the human body according to a normative ideal of what constitutes health, and medical interventions to remove or treat it would be appropriate. Scientists sometimes attempt to distance themselves from the evaluative meanings of a thick concept and use it purely as a factual descriptor. This effort is inherently misguided, however, since the factual and normative aspects are inextricably linked at the concept level. Dupré makes this point through an analysis of evolutionary biologists' attempts to investigate the causes of rape by studying the mating behaviors of flies and ducks. Since the concept of rape derives its meaning through a normative framework of consent, Dupré argues, attempting to use the concept as if it denotes a natural and timeless biological essence is a failure to grasp its meaning.[15]

13. Inmaculada de Melo-Martín and Kristen Intemann, "The Risk of Using Inductive Risk to Challenge the Value-Free Ideal," *Philosophy of Science* 83, no. 4 (2016): 500–520.

14. Bernard Williams, *Ethics and the Limits of Philosophy* (Abingdon, UK: Routledge, 2006).

15. Dupré, "Fact and Value," 34.

What are the practical implications of the use of thick concepts in scientific studies? On the one hand, the use of a thick concept does not necessarily affect the accuracy of the findings as long as scientists are committed to both its descriptive and normative content. On the other hand, if the findings are accepted as the factual background of subsequent practical deliberations about whether and how to act, this establishes the normative link between certain natural changes and a specific quality of badness without debate. Scientists, policy makers, or citizens who want to argue against policy action will either have to question the scientific basis for the claims, or maintain that the costs or side effects make it undesirable to act in response to them. But the underlying normative association between those particular natural changes and badness will stick. This gives scientists considerable power to set the basic normative terms for public debate on an issue.

During the scientific assessments for acid rain in the 1980s, the use of the thick concept of "damage" to describe changes in lake ecosystems due to acid deposition became the source of a controversy between scientists conducting assessments for the United States and Canada.[16] Although scientists broadly agreed on the changes observed in the lakes, they disagreed on the threshold of acidity at which these changes should be taken to indicate damage. US scientists took a more narrowly chemical analysis of the aquatic changes and argued that damage began at an acidity threshold of pH 5.0, while Canadian scientists took into account broader biological, geophysical, and human interactions, and maintained that damages started much earlier, at around pH 6.0. This semantic disagreement had important policy implications since it was understood that policy makers and the public would ignore any changes that were not defined as constituting damage. The definition of a concept thus has the power to make a natural event policy relevant—or not.

Hypotheses

Most phenomena that scientists study have myriad complex causes—proximate and distant, biological and social, micro and macro. In selecting hypotheses to test, scientists must narrow the universe of possible explanations and focus on a small number of plausible candidates. What constitutes a good explanation

---

16. Michael Oppenheimer, Naomi Oreskes, Dale Jamieson, Keynyn Brysse, Jessica O'Reilly, Matthew Shindell, and Milena Wazeck, *Discerning Experts: The Practices of Scientific Assessment for Environmental Policy* (Chicago: University of Chicago Press, 2019), 44.

and which variables are relevant to it depend on a combination of epistemic and practical factors. Empirical observations and background knowledge restrict the hypotheses that will be regarded prima facie plausible, but leave the choice open. Scientists must use their experience, judgment, and intuition to choose the hypotheses they will pursue. They might select from existing hypotheses in their field, or follow their imagination. The nonscientific, even nonrational processes involved in the generation of hypotheses are not taken to be of interest from a scientific point of view because hypotheses must always be tested against the evidence. Any influence of values would be eliminated through confirmation. The widely used distinction between the context of discovery, where values can be given free reign, and the context of justification, where they must be kept out, rests on this idea.

The factors that influence the formulation of hypotheses are not usually discussed and scrutinized. Scientists do not explain which possibilities they overlooked and why. Of course, information about absences is necessarily elusive. The best scientists could do is to show that they entertained a sufficiently diverse range of plausible explanations. The problem is that what constitutes sufficiently diverse and plausible also depends on the shared values and assumptions of scientists. Thomas Kuhn argued that most hypotheses that scientists entertain are suggested by their scientific paradigm.[17] Ordinary scientific activity consists in scientists working to extend an existing paradigm in predictable directions by testing the increasingly narrow and trivial hypotheses derived from earlier findings. This implies that the paradigm could prevent the emergence of entirely different ideas because it discourages their pursuit and also because existing knowledge determines the possibilities for new knowledge, while closing off some possibilities. Scientific discovery is path dependent.

While Kuhn assumed that new hypotheses would be derived from the existing body of confirmed ones, feminist philosophers of science have shown that hypotheses are determined by social factors too.[18] If most scientists working in an area share the same background beliefs or preferences about what is an important, interesting, or elegant explanation, the community as a whole

17. Thomas Kuhn, *The Structure of Scientific Revolutions* (1962; repr., Chicago: University of Chicago Press, 2012), 23–25.

18. Longino, *Science as Social Knowledge*; Elizabeth Anderson, "Feminist Epistemology: An Interpretation and a Defense," *Hypatia* 10, no. 3 (1995): 50–84; Elisabeth Lloyd, *The Case of the Female Orgasm: Bias in the Science of Evolution* (Cambridge, MA: Harvard University Press, 2009).

can miss significant ideas that don't fit its expectations, even while it continues to confirm many trivial hypotheses that enjoy some empirical support. Feminist scholar Evelyn Fox Keller made this point in explaining geneticist Barbara McClintock's success. She argued that McClintock's perspective as a woman and outsider in a male-dominated scientific community gave her the resources to devise an entirely novel kind of biological model—an interactive rather than hierarchical one—that would not have been possible had she shared the dominant perspective in her field that nature is organized hierarchically.[19]

The hypotheses devised and tested in a scientific community shape the possibilities for action at the stage of use. It is not enough for users to know that a finding enjoys some empirical support. They also need a sense of how complete the explanation is and how it compares with alternative explanations. An accurate explanation could be trivial, partial, or irrelevant for most practical purposes.[20] Many theories enjoy some evidentiary support, but offer only an incomplete explanation of a phenomenon. They perform well in some practical contexts and not in others. The problem is that the explanations that happen to be available acquire disproportionate practical significance if alternatives that nonscientists would consider more significant have not been pursued. For instance, if the only scientific study available on a question shows that a factor is *a cause* of an outcome, policy makers who want to take some action on the issue will not have much choice besides focusing on that cause and its implications, even if they do not consider it a particularly significant cause. After all, policy interventions will be more successful if they rely on evidence from scientific research, however partial and incomplete. But this diverts attention from potentially more important explanations that have not been studied scientifically. Decisions and policies based on science will be skewed toward actions that appear justified on the basis of the available explanations. Since the availability of an explanation is not just due to its scientific superiority but also to the values, concerns, and needs that led scientists to pursue it, following the science might mislead, and obscure full understanding of a problem, even while it facilitates certain forms of technical intervention.

The shift in medical research from the assumption of sameness to inclusion and difference, mentioned earlier, illustrates the practical implications of this

19. Evelyn Fox Keller, *A Feeling for the Organism: The Life and Work of Barbara McClintock* (San Francisco: W. H. Freeman and Company, 1983). See also Miriam Solomon, "Standpoint and Creativity," *Hypatia* 24, no. 4 (2009): 226–37.

20. Anderson, "Feminist Epistemology."

discussion. As Epstein points out, increasing attention to the biological bases of difference in medical research meant that other factors that could explain the large inequalities in health outcomes among different groups, such as those at the level of individuals, class, occupation, or nationality, were neglected. Since researchers increasingly tested and confirmed hypotheses that focused on biological differences, policy makers who wanted to improve health outcomes directed their attention to these differences. The explanations considered by a scientific community thus shaped collective understandings of physical and biological processes, and the kinds of policy actions that were considered feasible or desirable—or considered at all.

The controversy over GMOs provides another illustration of this dynamic. A common framing of this issue claims that there is no scientific evidence of harms from GMOs and those who resist the use of GMOs are denying sound science. Contra this widespread view, Hugh Lacey argues that opponents resist GMOs not because they are skeptical of science but rather because they have a different understanding of the kinds of scientific explanations that would be adequate to establish the claim that GMOs are safe. (Note the use of the thick concept "safe" here.) Through a case study of the resistance of popular rural movements for agroecology in Brazil, Lacey shows that these communities do not claim that existing studies are false but instead that they are partial and incomplete.[21] Studies that conclude that GMOs pose no risk to human health or the environment take a narrowly molecular approach to safety. They study the micro level chemical and biological effects of seeds in isolation from the context in which they are meant to be used, and find no cause for concern. The agroecology movement, by contrast, is concerned with the effects of GMOs in particular ecological and agricultural contexts. Its main worries are about the sustainability of GMO seeds and their effects on local biodiversity, global food security, and the self-sufficiency of small-scale farms. These effects are longer term, and at the level of the ecosystem and socioeconomic context; some effects would arise if farmers became dependent on commercial seeds protected by a global regime of intellectual property rights. This problem is exacerbated by the fact that GMO producers sponsor molecular studies that provide evidence for the

---

21. Hugh Lacey, "A View of Scientific Methodology as a Source of Ignorance in Controversies about Genetically Engineered Crops," in *Science and the Production of Ignorance: When the Quest for Knowledge Is Thwarted*, ed. Janet Kourany and Martin Carrier (Cambridge, MA: MIT Press, 2020), 245–70. See also Daniel Sarewitz, "How Science Makes Environmental Controversies Worse," *Environmental Science and Policy* 7 (2004): 385–403."

absence of gene-level toxicity, while longer-term contextual studies that would provide a picture of their full impacts are both expensive and underfunded. The result is that a gradually accumulating number of existing studies suggest to policy makers that the use of GMOs is safe or even beneficial, while studies of hypotheses about the long-run effects of GMOs in particular local contexts are largely absent. The available funding for scientific studies plays a crucial role in determining the kinds of explanations that the scientific community considers significant. I take up this issue in more detail in chapter 5.

Models

Many areas of modern science use numerical simulation models to analyze large-scale, complex physical processes such as the behavior of the earth's climate, spread of diseases, and decay of toxic contaminants. Models simplify features of the actual system to render it more workable, especially where using the actual numbers would be too complicated, or the underlying data are unavailable or incomplete. Scientists exercise discretion over the reasonableness of these simplifying choices, and the judgment of reasonableness is relative to the scientist's purposes in constructing the model. The evidence cannot establish the accuracy of the assumptions underlying a model, and approximations are false by definition. Scientists must rely on background scientific knowledge or common sense to guide their choices. Inaccuracy in background assumptions is often by design in order to ensure simplicity and practicality. Hence, there is a trade-off between the accuracy of a model and other features of it that are seen to be desirable. Insofar as the assumptions and approximations are explicit, models can provide useful information about the behavior of a system under different conditions.

The role of background assumptions and approximations in models, however, creates challenges for ascertaining the reliability of a model for use in a particular context. Since a model is a representation of a physical system, users naturally expect the model to be an accurate representation of the actual system. The trouble is that this expectation is impossible to meet.[22] To assess accuracy, the model's predictions must be compared with empirical observations. If observations contradict the predictions of the model, scientists can conclude that

---

22. Naomi Oreskes, Kristin Shrader-Frechette, and Kenneth Belitz, "Verification, Validation, and Confirmation of Numerical Models in the Earth Sciences," *Science* 263, no. 5147 (February 1994): 641–46.

something must be wrong somewhere, but the evidence cannot indicate where. Scientists can modify assumptions, change the model, or conclude that the observation is faulty. The evidence cannot determine or verify this choice. If tweaks to the model result in an improved fit between the model and data, this does not mean that the model has become a more accurate representation of the actual system. It only means that there is now increased consistency between the predictions of the model and the underlying system.

In fact, when the model is confirmed by observational evidence, the theoretical difficulties are in a sense greater.[23] To be able to deduce from this that the model is accurate, the scientist would need to know that the background assumptions and input parameters were all correct. If this could be established, then it would be possible to deduce that the model's accuracy must be responsible for the good predictions. But the correctness of assumptions cannot be tested, and approximations are inaccurate by definition. This means that it will be impossible to know why the model's predictions match the data. Confirmation can simply mean that some faulty background assumptions and approximations have canceled each other out. If two or more models are equally well supported by the available evidence, it is impossible to know which is the more accurate representation of the real system, just as it is impossible to know that a model that has not yet been constructed would not outperform all existing ones. The most one can say about the confirmation of a model is that it is adequate for a particular predictive purpose, such as predicting particular variables on a particular timescale. Even then, it cannot be ascertained whether superior performance results from the correctness of the underlying assumptions or sheer luck.[24]

For models to be useful in practice, policy preferences must be aligned with the predictive strengths of a model. Any attempts to use the model more broadly—for instance, as a representation of the system as a whole—will lack philosophical warrant. This significantly limits the practical uses of models and the possible courses of action that can be pursued on the basis of the information they provide. Policy makers would like to have reliable information about the behavior of a system and the subparts that are of interest to them, whereas scientists can only deliver models that cannot be verified and are adequate for certain narrow purposes.

23. Oreskes, Shrader-Frechette, and Belitz, "Verification, Validation, and Confirmation of Numerical Models."

24. This resembles some of the difficulties with machine learning models, which are likewise obscure in their workings and can only be judged by their predictive success.

The impossibility of verifying models empirically means that models will be skewed toward values and purposes assumed by the scientist. Model construction requires many choices that cannot be dictated by the best available understanding of the underlying scientific mechanisms. Decisions about how to fix inputs for different parameters, how much to simplify different subparts, which natural processes to leave out, or which approximations to make cannot be determined on the basis of scientific knowledge. They must be made with reference to an assumed purpose.[25] These decisions will play an important role in determining the strengths and weaknesses of the model; the model is more likely to be adequate for purposes that scientists have deemed more significant. This, in turn, determines the kinds of policies that the model can inform. While I showed that ordinary scientific hypotheses are also shaped by the interests and priorities of scientists, the processes of testing and confirmation in scientific inquiry ensure that hypotheses are ultimately accepted or rejected on the basis of the evidence. The impossibility of verifying models rules out this possibility, which means that the underlying value judgments cannot even be fully checked against the data.

Modeling decisions are innumerable, opaque, and often forgotten and buried beneath many iterations of the model over time. Scientists are guided broadly by what they consider most significant with respect to an assumed purpose. This can be a social or political purpose—for instance, based on how scientists imagine policy makers might want to use their model—or it can be the scientist's own professional or personal purposes. These will determine which scientific processes and effects are considered important to represent more accurately, and which ones are omitted or approximated roughly. Judgments about significance are, of course, entirely subjective. Since scientists do not have access to the values and priorities of others at the research stage, they will either rely on their personal preferences or try to think about what would be considered significant by others.

Geographers Martin Mahony and Mike Hulme's study of the establishment of the United Kingdom's Hadley Center for climate research illustrates how modeling choices that appeared to be purely scientific actually involved deep value conflicts about national and global priorities.[26] In the early days of the

---

25. Wendy Parker and Eric Winsberg, "Values and Evidence: How Models Make a Difference," *European Journal for Philosophy of Science* 8, no. 1 (2018): 125–42.

26. Martin Mahony and Mike Hulme, "Modelling and the Nation: Institutionalising Climate Prediction in the UK, 1988–92," *Minerva* 54, no. 4 (2016): 445–70.

center, British climate modelers faced competing pressures. On the one hand, they wanted to prioritize national policy goals, and felt the pressure to compete with their US and German counterparts, who had developed the leading climate models of the day. On the other hand, they aspired to make British models the leaders in collaborative European and global efforts to tackle climate change, which would mean sacrificing national priorities to aim for models with more global strengths. This also shows that national scientific communities may develop collective preferences over values and face collective dilemmas over modeling trade-offs.

Modeling decisions can be made with reference to purely professional purposes too. Scientist can make decisions based on what would make the model more productive for future research, easier to work with, or more relevant for their own research agenda. Practices may vary across disciplines; a physical scientist and biologist will value different features of a climate model. Sometimes these decisions will be made simply on the basis of what is possible or feasible. For example, the current state of climate modeling allows the construction of models that are successful at predicting global mean surface temperatures, but not regional variables such as precipitation, wind speed, and sea-level pressure. This is one reason why climate policy debates have for a long time focused narrowly on preventing global temperature increases rather than, say, considering local adaptation measures.

Scientists' judgments about significance determine the kinds of mistakes they want their model to avoid. They can decide whether it is better for the model to err on the side of underestimating or overestimating certain effects, and what kinds of mistakes it would be fine to tolerate and which ones must be avoided. The choice can be made with an eye to the political or distributive consequences of different kinds of mistakes, or the professional costs. Preferences about the magnitude, cost, and distribution of mistakes will result in models that are more accurate for certain regions or timescales, and with respect to different natural processes. This, in turn, will determine which political problems and needs can be addressed more effectively with existing models.

Evidence

In designing a study, scientists must decide on the types and sources of evidence they will collect, and what they will accept as sufficiently good quality. This decision affects the likelihood of error in the judgment to accept or reject a hypothesis later on. Douglas has shown that the judgment that the evidence is sufficient is always relative to an assumed purpose: Sufficient for what?[27] Scientists must consider the consequences and decide based on how bad it would be to make different types of mistakes. Douglas's paradigm case is about the sufficiency of evidence to accept a hypothesis, but the issue that she detects arises at other stages in the research process as well. My aim is to demonstrate how disciplinary norms about the quality of evidence presuppose value judgments about the relative acceptability of different kinds of errors under uncertainty, and the consequences of action and inaction.

What constitutes good-quality evidence is usually determined by shared standards, which are established with reference to the scientific or practical goals of a discipline along with constraints on the available resources, data, and methodological training. Once standards become widely accepted, their instrumental justification vis-à-vis particular purposes may recede into the background. Adherence to these standards comes to define the idea of scientific excellence in a field, and shapes the professional identity and habits of practitioners. Although these standards are internal to a practice, they acquire external significance if the findings become relevant for policy decisions. When this happens, the values assumed in establishing standards of evidentiary quality affect the goals that can be pursued on the basis of the evidence, even though the standards have not been set with any particular policy goals in mind. This results in a mismatch between scientific and political purposes, which is disguised as a technical debate about whether the evidence is in fact good quality. When different scientific disciplines have different standards of evidentiary quality, the value orientations that have led to their adoption can be obscured by arguments that appeal only to scientific accuracy and rigor.

Philosopher Jonathan Fuller's discussion of two competing scientific approaches to evidence used during the COVID-19 pandemic provides a good illustration of what I mean.[28] The first approach that he identifies is typically associated with public health epidemiology and is more pluralist about evidentiary quality. It uses many different types and sources of evidence—lab

27. Douglas, *Science, Policy, and the Value-Free Ideal*.

28. The description of the two approaches is from Jonathan Fuller, "Models v. Evidence," *Boston Review*, May 5, 2020, http://bostonreview.net/science-nature/jonathan-fuller-models-v-evidence. For more on the first approach, see Julian Reiss, "Pragmatist Theory of Evidence," *Philosophy of Science* 82, no. 3 (2015): 341–62. For further discussion of their strengths and weaknesses, see Miriam Solomon, *Making Medical Knowledge* (New York: Oxford University Press, 2015).

based, population based, or clinical—and accepts data of varying quality. It is also open to diverse methodologies and tools, relying on information derived from models and theory alongside empirical data. Public health epidemiologists build many different models that use different techniques, background assumptions, and parameters. Since it is recognized that both the evidence and techniques used may be subject to biases, this approach typically produces a range of divergent predictions, whose accuracy cannot be fully ascertained. This problem was evident in a highly publicized disagreement between the disease models from Imperial College and Oxford early on in the COVID-19 pandemic; I discuss this in chapter 8. Still, this approach can generate predictions that are successful on certain scales and give estimates of worst- and best-case scenarios. It may be difficult to obtain results that enjoy strong scientific warrant, but findings can still be informative and possibly good enough, depending on one's purposes.

Fuller argues that a second approach, associated with clinical epidemiology, holds evidence to a higher standard—namely to the gold standard of a well-conducted randomized experiment. Evidence from experiments is considered to be intrinsically more reliable than observational evidence because it can eliminate most sources of bias. Alternative sources of evidence are accepted or rejected on the basis of how closely they can approximate this standard. A greater proportion of data is regarded as low quality and discarded. Scientific disciplines that take this more demanding view are willing to accept a lot less evidence as good quality and therefore are more likely to reject many interventions as not based on good evidence.

My aim is not to assess the strengths and weaknesses of these approaches, or offer one correct understanding of evidentiary quality, but rather to emphasize how the adoption of different notions of quality follows from judgments about values and priorities. What appears on the surface to be a purely scientific difference about rigor conceals normative differences between these disciplines over how best to further public health. The first approach is more inclined toward action and ready to recommend precautionary interventions against diseases. Since models can produce adequate predictions without offering new understandings of the behavior of a system, an approach that relies on modeling will have an inherent bent toward intervention over understanding. This approach will also be more tolerant of false positives, with a greater likelihood of confirming inaccurate findings.

The second, experimental approach, by contrast, reflects a more skeptical disposition and an inclination toward caution. It is more worried about inflicting harm through unjustified interventions than in allowing a bad status quo to persist through inaction. This tendency becomes more pronounced as the possibility of acquiring sufficiently high-quality evidence in an area becomes more difficult. While proponents of this view might insist that their belief in the importance of scientific rigor drives their demand that policy interventions be backed by "evidence"—usually shorthand for evidence from randomized control trials—it is equally possible that a background aversion to intervention, at least to faulty intervention, drives their scientific standards higher. The exact causal relationship between standards of evidentiary quality and background orientations toward action may be impossible to determine, but there is a clear and necessary affinity between the two, which becomes particularly pronounced at the point of decision-making.

Different disciplinary approaches to evidentiary quality incorporate biases toward one or the other type of error, which map onto different stances toward and interpretations of precautionary action. After all, what counts as precautionary is inherently subjective. It depends on the outcomes that an agent wants to avoid most, especially in cases where both action and inaction could lead to bad outcomes. A historical and sociological study of scientific disciplines could help excavate the factors that led different disciplines to develop different understandings of precaution. The medical profession's long and checkered history of interventions that did more harm than good might justify its continued emphasis on the mantra "first, do no harm."[29] Meanwhile, public health has faced the pressure of devising protective measures in the face of serious challenges to collecting high-quality population-level health data. This has meant that it had to devise a more flexible, permissive, and pluralist approach to using any and all available knowledge.

There may be good reasons why disciplines have the standards that they do, although this too is always worth questioning. What is crucial for my purposes is to underscore that these reasons are always derived from the particular values assumed by practitioners. Considerations about the quality of evidence are never purely scientific; they cannot be resolved by appeal to truth-related criteria. Scientists themselves sometimes contribute to confusion over this point, such as when they reject different approaches to evidence for failing to meet criteria of scientific rigor rather than for embodying a different practical calculation about the consequences of action and inaction, or a different

29. David Wootton, *Bad Medicine: Doctors Doing Harm since Hippocrates* (New York: Oxford University Press, 2007).

philosophy about how best to achieve an underlying goal. Once we understand the affinity between approaches to evidentiary quality and normative orientations toward different goals and outcomes, it should become clear that discussions over evidence necessarily entail a position on errors and stakes, and should not proceed without attention to them, at least when the practical use of science is in question.

Randomized Control Trials

The final element of research design that I want to discuss is the composition of randomized control trials. I aim to show, as in the last section, that what appear to be purely scientific views about best practice in trial composition actually presuppose normative views about the right way to address a problem. The right way to design trials became the source of a controversy during the AIDS crisis, when activists contested the widely accepted scientific approach. Scientists believed that the correct composition for a trial should be decided on purely scientific grounds. The accepted view for AIDS trials was that homogeneous populations—excluding Blacks, women, intravenous drug users, and patients taking other medications—would constitute the ideal trial group because they would minimize possible biases and confounders.[30] AIDS activists rejected this view and argued that what constituted a good study population had to be defined relative to the practical purposes of the study. Homogeneous populations and subjects who had no other medical problems would not be appropriate stand-ins for real patient populations, which did not resemble the homogeneous groups preferred by scientists.

Epstein documents how the AIDS movement succeeded in changing the scientific attitude toward randomized control trials. Scientists began to recognize two scientifically valid approaches to the design of trials.[31] The first, "pragmatic" approach prioritizes the practical purpose of solving problems in clinical practice.[32] Homogeneous populations with patients taking only the drug under study are not the ideal population for these studies because these groups

---

30. Steven Epstein, *Impure Science: AIDS, Activism, and the Politics of Knowledge* (Berkeley: University of California Press, 1996).

31. Epstein, *Impure Science.*

32. See Alvin Feinstein, "An Additional Basic Science for Clinical Medicine: II. The Limitations of Randomized Trials," *Annals of Internal Medicine* 99, no. 4 (1983): 544–50; Robert J. Levine, *Ethics and Regulation of Clinical Research* (Baltimore: Urban and Schwarzenberg, 1986).

are not suitably analogous to the target patient population. Real patients have different conditions and proclivities, and take many medications at the same time to increase their chances of survival. A drug tested on a pure trial population may therefore not produce the expected effect in real patients. The second, "fastidious" approach is the more conventional one. Here the purpose of the trial is to understand as clearly as possible what effects the drug produces in the human body. This requires an artificially controlled, homogeneous trial group and yields less ambiguous, more secure findings. The weakness of this approach is that the tested drug may not produce the same results in real clinical subjects.

The recognition of two different approaches makes it possible to say in hindsight that scientists and public health officials should select the research design that fits their purposes. But what I want to stress is not just the fact that researchers and activists disagreed about the immediate purposes that should be adopted. The more interesting point is that the scientists had assumed that running trials in the way that they had recognized as constituting scientific best practice up until then—the fastidious approach—was the correct practical approach to saving lives too. This, however, was an unexamined background assumption about how best to act and its relationship to scientific beliefs about good research practice. The success of activists was not only in challenging the composition of trials to help a particular patient population but also in exposing the purpose relativity of norms around scientific best practice.

What Is the Problem?

I have described how scientific findings are shaped by the values, purposes, and background assumptions of scientists, starting from the earliest stages of research. These examples do not exhaust possible roles for values in scientific research, but they highlight ones that are both salient at the stage of use and unlikely to be detected by the mechanisms of self-correction within the scientific community. These values and purposes may not be problematic from a scientific perspective, and may even exemplify best practice in a field, and yet they can still have problematic implications at the stage of use.[33] I now want

33. For the use of randomized control trials in policy making, this point is emphasized in Nancy Cartwright and Jeremy Hardie, *Evidence-Based Policy: A Practical Guide to Doing It Better* (New York: Oxford University Press, 2012).

to say more about why exactly the influence of values and purposes on scientific research is problematic from a democratic perspective.

In brief, the main source of the problem is that science and scientists have their own values, purposes and standards, which shape their research, and these may not map onto values, purposes, and standards held by or acceptable to others who must rely on scientific findings to pursue different aims. This may be primarily because scientists prioritize the pursuit of shared professional aims that are significant within the community, or because their own partial perspective about what constitutes socially and politically significant goals informs their scientific choices and standards. These two things may well be impossible to disentangle in practice. Either way, values and purposes presupposed during research determine the ends that will be possible and desirable to pursue on the basis of this knowledge as well as the types of errors that are more likely to result from relying on scientific evidence under uncertainty.

Democratic ends and priorities are not formed prior to and independently from what the science says but rather in light of and sometimes in reaction to it. Scientific findings may introduce a new problem to the political agenda or offer possible solutions to existing problems. The available scientific knowledge shapes what citizens and policy makers take to be possible, feasible, or likely, which in turn plays a crucial role in guiding evaluative judgments about different courses of action and possible outcomes. In fact, this is usually taken to be a good way for politics to relate to scientific claims, if not the ideal way. The familiar division of labor model stipulates that democratic purposes should be shaped in light of the best available knowledge. It is common to deplore the fact that citizens and policy makers fail to form their purposes in light of scientific findings, but not so common to think that there might be something wrong with doing so.

The problem is that this model works only if science itself is free from presumed values and purposes, and simply reports on how things are. If scientific assessments incorporate judgments about what is significant, adequate, plausible, relevant, or useful, and political goals are determined in light of research that incorporates such judgments, the values and purposes of scientists will end up preempting or circumscribing democratic deliberation from the outset. Moreover, this influence will operate invisibly because science is taken as offering facts—uncertain and fallible to be sure, but still facts. Scientists themselves may be unaware of how their values have shaped their claims about facts and see their research choices as dictated purely by good scientific practice.

Since it is usually better to consult available knowledge rather than ignore it, policy makers and citizens focus their attention on issues that the available science can explain well, which then determines the policy options that are seen to be feasible.

The determination of feasibility, much like the determination of significance, adequacy, plausibility, relevance, and usefulness, is both subjective and deeply political.[34] While the laws of nature and limits of human capacities put some uncontroversial constraints on what is possible, a wide range of possible actions appear more or less feasible depending on what we know about them. If we know nothing about a certain option or our knowledge about it is highly uncertain, we cannot consider it feasible, but this is importantly different than knowing that it is not feasible. What we happen to know about a scientific issue that has become politically salient is never exogenous. It is determined by choices that researchers have made at stages of scientific inquiry that are rarely thought to be relevant to the political process. The feasibility of policies is thus constructed through these unexamined scientific decisions rather than revealed or discovered by them. The more uncertain the available scientific knowledge base, the more likely it is to reflect a partial perspective. Scientists' decisions about which hypotheses to pursue, evidence to collect, variables to study, or experiments to run will have greater influence on the shape of the results under greater uncertainty.

To translate these concerns from philosophical to political vocabulary, we can say that science and scientists hold a crucial power to set the agenda, frame the debate, and determine beliefs about feasibility. The challenge is to be precise about why this power is troubling from a democratic perspective and which democratic values are threatened by it. Some of the major theoretical frameworks through which the problem of expertise in democracy has been examined fail to offer the conceptual vocabulary to describe this concern. The issues that I have highlighted cannot, for instance, be expressed in terms of the common worry about expert domination, at least on the standard conception of domination as the capacity of one agent to interfere arbitrarily with another agent's choices.[35] Scientists do not usually possess such arbitrary power over

---

34. Holly Lawford-Smith, "Understanding Political Feasibility," *Journal of Political Philosophy* 21, no. 3 (2013): 243–59.

35. For this account of domination, see Philip Pettit, *On the People's Terms: A Republican Theory and Model of Democracy* (Cambridge: Cambridge University Press, 2012). For an analysis of the problem of expertise through the domination lens, see Henry S. Richardson, *Democratic*

other citizens. Nor do scientists possess de jure illegitimate decision powers since they are typically not delegated the power to make binding rules and laws. (They possess political influence as scientific advisers, but I leave that to the next chapter.) Of course, I am arguing that scientists do have the power to shape and influence the choices available to other agents, but this power operates through what they offer as the shared knowledge base on which collective decisions can draw. This is a structural form of power, and it cannot be captured by strictly agential accounts of domination or juridical analyses of power.[36]

Epistemic accounts of democracy, which focus on the truth of decisions, cannot fully capture this concern either.[37] It might seem at first that they would offer a promising framework since democracies turn to expertise because they want to base policies on reliable information and thereby improve outcomes. If expert-driven policy fails, it is reasonable to think that the failure would also be in terms of outcomes. Indeed, in some of the cases described above, the problem could be expressed in part as a bad outcome. For instance, the neglect of social group variables in medical research was partly responsible for the lagging health outcomes of women and minorities, and a dispute over the composition of randomized control trials will result in better outcomes for some and worse outcomes for others, whichever way it is resolved.

Insofar as epistemic accounts aim to evaluate decisions by their truth and institutions by their truth conduciveness, however, they will miss the crucial role that science plays in determining the standards of truth that we can rely on to evaluate democratic decisions and institutions. Scientific views about the nature of a problem and possible responses to it offer the evaluative resources for citizens to judge policy decisions. Things not known to be part of a problem and interventions not known to be possible will not be part of citizens' expectations for future good outcomes and their criticism of the status quo. If existing scientific models do not study the effects of a pandemic policy response besides a strict lockdown, citizens cannot say if the number of deaths associated with a lockdown is too many or too few. And a strict lockdown will

Autonomy: Public Reasoning about the Ends of Policy (New York: Oxford University Press, 2002); Sabeel Rahman, *Democracy against Domination* (New York: Oxford University Press, 2016).

36. On structural power, see, for example, Iris Marion Young, *Justice and the Politics of Difference* (Princeton, NJ: Princeton University Press, 1990); Sharon Krause, *Freedom beyond Sovereignty: Reconstructing Liberal Individualism* (Chicago: University of Chicago Press, 2015).

37. See, for example, Cathrine Holst and Anders Molander, "Epistemic Democracy and the Role of Experts," *Contemporary Political Theory* 18, no. 4 (2019): 541–61.

be considered a better outcome than doing nothing at all, if only because scientists have not modeled other possibilities. Findings might be accurate and yet still partial, and therefore misleading.

When the best available scientific evidence guides the assessment of the truth of decisions, citizens' ability to make certain kinds of criticism is blocked, just as their ability to make others is enhanced. Scientific knowledge opens up some lines of thought while obscuring others. Even if citizens can step outside existing problem framings to resist so-called evidence-based policies that go against their needs and values, their resistance might be dismissed as groundless and unscientific because the science that would have supported their actions has not been conducted. The conflict between the Brazilian agroecology movement and biotech companies over the safety of GMOs exemplifies this problem. How the existing science limits the possibility of democratic criticism and the evaluation of policies will not be visible unless the science is questioned and criticized first.

The democratic problem with the role of science and scientists is best expressed in terms of equality. Democracy requires citizens to have an equal opportunity to influence and contest the decisions to which they are subject. But if the values and aims presupposed in scientific research shape decisions, then an important determinant of the policy process is shielded from citizen input and scrutiny. The influence that citizens may have over the selection of aims *in light of* the knowledge presented is circumscribed at an earlier stage through what is offered as the neutral factual background. This means that some citizens—scientists and those who influence them, such as through funding decisions—have the opportunity to determine the shape and bounds of policies not only through regular channels of political participation but also through scientific decisions not recognized as having political relevance. This creates an inequality in opportunities for political influence. As such, it is both intrinsically problematic and likely to lead to substantive inequalities in the resulting policies, as the availability of good-quality scientific findings determines whose needs and concerns can be met more effectively, favoring some citizens over others.

The idea that expertise threatens the equality fundamental to democracy is not a new thought, but it is usually discussed in the context of the bureaucratic delegation of power to experts.[38] The standard division of labor between experts

38. See Robert A. Dahl, *Controlling Nuclear Weapons: Democracy versus Guardianship* (Syracuse, NY: Syracuse University Press, 1985); Stephen Turner, "What Is the Problem with

and laypeople, in which the latter define the aims of society and the former supply the means for attaining these ends, is offered as a solution to this problem.[39] My aim is to reveal the limits of this model by focusing on the production of knowledge itself. If scientific knowledge is shaped by some set of assumptions about ends and democracy requires citizens to have a say over ends, then it follows that the public must have a meaningful chance to engage with the science and examine the ends that scientists have presupposed. This means that the worry that gave rise to the means-ends division of labor solution—namely that democratic policy would be guided subtly by the priorities of experts—cannot in fact be addressed through the division of labor itself.

Of course, the public and politicians always retain the power to reject scientific knowledge tout court, and exercise this power often enough. It may thus seem that whatever power science has in politics is easily thwarted by dismissal and denial. But rejecting scientific knowledge is a blunt instrument that causes more severe damage in most cases. The options of complete acceptance or denial are poor substitutes for the power to effectively examine and challenge scientific assumptions to make them more fit to serve democratic aims. This is not because ordinary citizens are presumed to have a superior claim to scientific competence but instead on the democratic grounds that their needs and priorities should be represented in democratic decisions.

Not all theories of democracy will find it problematic that scientists have the agenda setting and framing powers that I have described. Minimalist accounts of democracy, which hold that democratic equality can be satisfied through universal suffrage, and do not require further attempts to equalize opportunities for participation and influence, will not be troubled. Elitist theories, which maintain that it is the responsibility of elites to set the agenda and determine the ends to be pursued, might even regard this as an ideal. I simply have a more egalitarian and participatory conception of democracy than these accounts. Even among those who believe that the public should be empowered to participate and influence decisions between elections, many these days will wish that scientists had more power and influence. This is an understandable response to the widespread rejection of science and bad outcomes that have resulted from it. My argument, though, is not about the amount of power that scientists should have but rather what it would take to make their

---

Experts?," *Social Studies of Science* 31, no. 1 (2001): 123–49; Thomas Christiano, *The Rule of the Many: Fundamental Issues in Democratic Theory* (Boulder, CO: Westview Press, 1996).

39. For a defense of this view, see Christiano, *The Rule of the Many*.

opportunities for political influence compatible with equality among citizens in a democratic society. Once we recognize the role that scientists' own values play in the science they produce, and the role this plays in determining which problems can be addressed and how, it is not clear what, besides elitism, could justify the belief that scientists' aims should be everyone else's too.

What Is to Be Done?

The point of starting from an account of the role of values in science was to be in a better position to offer a different model for how to use science in a democracy. Understanding the sources of possible failures is the first step in devising procedures and institutions to address them. My suggestions will thus follow closely from the account developed so far. Before turning to suggestions, however, I want to show how the discussion so far helps rule out some widely held views about the proper relationship between science and democracy.

I have already pointed out that the standard division of labor model, which stipulates that the public and policy makers choose their ends in light of the facts provided by experts, cannot be maintained because it rests on the assumption that expertise can be taken as providing value-free knowledge. Once we understand how scientists' values and purposes shape their research—from the way they define concepts to the hypotheses they test, models they build, and evidence they accept—we cannot assume that science simply delivers an account of things as they are. Or more precisely, it offers one partial picture of the way things are, which omits and obscures even while it explains and illuminates. This conclusion undermines the division of labor model. Strict adherence to this model—for example, by asking scientific advisory bodies to produce a summary of how the science stands and then accepting those conclusions as the factual basis of policy deliberations—would produce policies shaped imperceptibly by the assumptions and purposes of scientists.[40]

40. Although there have been other critiques of this model (most prominently by Sheila Jasanoff in, for example, *The Fifth Branch: Science Advisers as Policymakers* [Cambridge, MA: Harvard University Press, 1990]), it is nonetheless frequently adopted in practice. Boris Hauray and Philippe Urfalino show how medicine licensing by the European Union involves precisely such a process where policy makers take the scientific assessment without question from expert committees. Boris Hauray and Philippe Urfalino, "Mutual Transformation and the Development of European Policy Spaces: The Case of Medicines Licensing," *Journal of European Public Policy* 16, no. 3 (2009): 431–49. Roger Pielke Jr. similarly argues that the idea is alive and well

Another approach ruled out by this analysis is to aim at reducing uncertainty through further research and scientific debate. For one thing, it is not clear that more research or debate would reduce uncertainty instead of uncovering new and deeper sources of it.[41] Moreover, any further research will be subject to the same difficulties with the role of background assumptions and values. These are endemic problems, not incidental ones. In a sense, they are due to the fundamental uncertainty and incompleteness of science. If scientists could attain the certainties of logical deduction and the completeness imagined in some unified visions of science, perhaps there would be no reason to worry about omissions and bias. But it is not particularly meaningful to target uncertainty and incompleteness as a response since we cannot even conceptualize what certain and complete science would look like, let alone achieve it in practice.[42] The real problem is that the way that scientists respond to uncertainty and incompleteness through methodological and theoretical choices yields knowledge that is skewed toward values and purposes they have considered significant, rather than distributed randomly or evenly. Detecting and addressing this issue should be the main target of efforts to improve the use of science.

Douglas argues that recognizing the impossibility of value-free science places an ethical obligation on scientists to consider the social consequences of their work. Scientists have a duty to use social and political values to guide their choices at points in the research process where value judgments are needed.[43] Scientists must imagine possible uses of their research, decide which consequences are more important to bring about or which mistakes better to avoid, and use these judgments in scientific choices not determined by the evidence. For instance, a climate modeler should think about the ethical implications of likely inaccuracies on different timescales and geographic

---

among policy makers in the United States. See Roger Pielke Jr., *The Honest Broker: Making Sense of Science in Policy and Politics* (Cambridge: Cambridge University Press, 2007).

41. Jasanoff, *The Fifth Branch*, 8; Jerome R. Ravetz, "Usable Knowledge, Usable Ignorance: Incomplete Science with Policy Implications," *Knowledge* 9, no. 1 (1987): 87–116.

42. Nancy Cartwright, *The Dappled World: A Study of the Boundaries of Science* (Cambridge: Cambridge University Press, 1999).

43. Douglas, *Science, Policy, and the Value-Free Ideal*; Heather Douglas, "The Moral Responsibilities of Scientists (Tensions between Autonomy and Responsibility)," *American Philosophical Quarterly* 40, no. 1 (2003): 59–68. For a response, see S. Andrew Schroeder, "Using Democratic Values in Science: An Objection and (Partial) Response," *Philosophy of Science* 84, no. 5 (2017): 1044–54.

locations, and ensure that modeling choices reflect the knowledge needs of different populations.

I will discuss the merits of this proposal for scientists in their capacity as policy advisers in the next chapter. For now, I want to focus on scientists making such judgments in their role as researchers. The appeal of Douglas's suggestion lies in the intuition that it would be good—that is, both useful and morally desirable—if the science produced in a society were more consciously designed to advance the moral and political values of that society, rather than scientific or random ones. Yet questions such as what kinds of mistakes would be worse or which timescales or geographic locations are more significant are thoroughly political and highly contested. Expecting scientists to discern and use social and political values in their research would be to assign scientists a duty of political representation. This is a role for which they are neither qualified nor properly authorized. The task of representing public values is delegated to political representatives through elections. Scientists informally claiming this responsibility out of a sense of professional or personal duty would not only lack a legitimate formal basis but also would be undesirable. Scientific qualifications are not relevant to making political judgments in the name of other people, and scientists cannot know what public values and preferences are, especially since there is usually disagreement or ambiguity around these values prior to scientific assessments. There is no good epistemic basis on which they might take guesses, and in fact it would be wrong to conceptualize democratic values and preferences as preexisting facts to be guessed or discovered. Values and preferences are formed through processes of contestation, deliberation, and compromise in the public sphere.

To implement Douglas's proposal, scientists would fall back on consulting their own moral judgments about what constitutes the public good, styling themselves as trustee representatives. In fact, I suggested that this might already be how scientists make some decisions that require judgments about significance, values, and priorities. This would mean that scientists' own judgments about social and political matters would play an important role in shaping the available options on scientific questions. Furthermore, this influence would operate not through explicit advocacy but instead through the way that the science itself is conducted. This is precisely what I have identified as worrisome from a democratic perspective. This is the problem, not a solution. If scientists relied on moral judgments in their decision to accept a hypothesis, those who disagree with their values would have reason to distrust their scientific conclusions. This would result in a loss of information, and blur the useful distinction between scientific standards and political ones. Those who used the findings would not able to determine how using other values would have changed the conclusions. This would be problematic both epistemically and democratically.[44]

I will say more about democratic input in setting priorities for funding science in chapters 5 and 6, but it should not be controversial to claim that if clear and widely shared democratic priorities exist, and are widely known, these should be used to guide publicly funded research. Scientific issues on which there are clear and widely shared democratic priorities are usually not the ones that become politically contentious, though.[45] The difficult cases are those in which there is either no clear democratic will or disagreement over it, and democratic decisions depend at least in part on answers to scientific questions. In these cases, it is not desirable for scientific research to incorporate speculative assumptions about democratic priorities at earlier stages of research. The only way to ensure that policies are not determined by the purposes assumed by researchers is to submit scientific findings to scrutiny at the decision stage.[46] This would prevent the preemption of democratic deliberation and decision-making by scientific findings that are imperceptibly value driven.

Democratic scrutiny should be directed toward exposing the values and assumptions of science to ensure that they do not restrict democratic decision-making to narrow, partial goals. It should also aim to encourage scientists to recognize and possibly revise their assumptions, thereby increasing democratic influence over the agenda and framing powers of science without forgoing the benefits from knowledge. Again, the aim is not to make science either more or less powerful but instead to open up its power to wider discussion and contestation. This process should be democratic rather than restricted to members of the scientific community because questions about the appropriate aims to pursue, errors to avoid, and needs to prioritize are fundamentally political. Such scrutiny can take place in a variety of settings, including public debates between scientists, congressional or committee hearings, local town

44. Douglas also proposes that scientists could rely on deliberative minipublics to discern the values of the public. I discuss my objections to this solution in chapter 4.

45. Yaron Ezrahi has a helpful two-by-two typology of issues according to scientific versus political consensus versus dissensus. See Yaron Ezrahi, "Utopian and Pragmatic Rationalism: The Political Context of Scientific Advice," *Minerva* 18, no. 1 (1980): 111–31.

46. The need for critical scrutiny of expertise is also emphasized in Alfred Moore, *Critical Elitism: Deliberation, Democracy, and the Problem of Expertise* (Cambridge: Cambridge University Press, 2017).

hall meetings, or the media. It can involve participation by different actors, from scientists, policy makers, and legislators to nongovernmental organizations (NGOs), stakeholders, and ordinary citizens. Social movements have often subjected science to critical scrutiny, challenging or resisting its findings and methods.[47] On issues from AIDS to women's health to environmental preservation, activists have confronted experts squarely on scientific territory, leading them to revise their approaches to align more closely with the needs and interests of other people.[48] The successes of at least some social movements in this regard shows that the democratic scrutiny of science is possible, even on complex scientific issues. The following two chapters consider how this contestatory spirit may be institutionalized in ordinary scientific advisory processes so that it becomes an essential part of how science is handled in democracies. Here I will schematically describe the basic features of this process.

The democratic examination of expertise can be conceptualized as having three main stages.[49] The first is diagnosis, aimed at identifying possible biases in knowledge claims, especially focusing on the role of assumptions, values, and omissions. This is an essentially critical task that could involve philosophical critique too. The discussion in this chapter, for instance, could be interpreted as contributing to a diagnostic effort by pinpointing potential sources of problems at a general level, although each issue also requires specific attention. The diagnosis process can involve scientists challenging each other, or activists, politicians, and other citizens challenging scientists. While scientists' superior understanding of the science may make them more qualified to scrutinize the assumptions driving each other's work, assumptions shared by scientists working in an area might be more effectively detected and exposed by outsiders. The medical paradigm shift from sameness to inclusion and difference, the agroecology challenge to GMOs, and AIDS activists' challenge to drug trial compositions all exemplify this.

The diagnostic task is likely to be more effective if different sources of knowledge can be compared. The more diverse the scientific approaches to a problem, the easier it will be to detect their assumptions and examine their

47. Moore, *Critical Elitism*, 95–117.

48. Epstein, *Impure Science;* Wendy Kline, *Bodies of Knowledge: Sexuality, Reproduction, and Women's Health in the Second Wave* (Chicago: University of Chicago Press, 2010).

49. This section was inspired by and loosely follows the structure in David Wiens, "Prescribing Institutions without Ideal Theory," *Journal of Political Philosophy* 20, no. 1 (2012): 45–70.

strengths. It will also be easier to find competent experts to criticize each other, thus giving the public and policy makers a clearer view of the limitations of each. Proponents of the public health and clinical epidemiology approaches to evidentiary quality actually debated one another through mainstream publications during the COVID-19 pandemic, thereby clarifying the aims of each for a public audience.[50] Such organized debate does not usually happen in full public view, so careful institutional design can be necessary to facilitate similar encounters, and make matters of expertise accessible and intelligible to the public. I will develop this idea in greater detail in chapter 4.

The second phase of scrutiny involves identifying and using relevant corrective information that might improve the science, or the fit between science and democratic purposes. This can involve local or experiential knowledge that challenges the background assumptions of scientists, or points out a missing perspective regarded as significant by nonexperts, but that was not considered in the scientific literature. New information can improve existing studies, or it might simply show that they are wrong, irrelevant, or not reliable enough. Even if it is not possible to improve or redo studies, knowing the limits of the available knowledge is crucial for allowing policy makers and the public to calibrate their confidence in the accuracy and usefulness of rival claims. The identification of omissions constitutes valuable information in itself about the completeness and reliability of existing explanations.

To be clear, I am not making the strong empirical claim that democratic scrutiny will improve the accuracy of science. My argument is therefore different than those of epistemic democrats, who claim that democratic processes are in some circumstances more likely to arrive at the truth than small groups of experts.[51] My point is rather that since scientific claims face the limitations that I have discussed, it is important for democratic processes to focus on identifying and responding to them, as opposed to proceeding on the basis of science that favors some options over others. Democratic scrutiny may or may not generate knowledge that ultimately improves the accuracy of

50. Marc Lipsitch, "Good Science Is Good Science," *Boston Review*, May 12, 2020, http://bostonreview.net/science-nature/marc-lipsitch-good-science-good-science; John P. A. Ioannidis, "The Totality of the Evidence," *Boston Review*, May 26, 2020, http://bostonreview.net/science-nature/john-p-ioannidis-totality-evidence."

51. Hélène Landemore, *Democratic Reason: Politics, Collective Intelligence, and the Rule of the Many* (Princeton, NJ: Princeton University Press, 2008); Robert Goodin and Kai Spiekermann, *An Epistemic Theory of Democracy* (Oxford: Oxford University Press, 2018).

science; its primary purpose is to produce policies that are more responsive to citizens' needs and avoid certain bad outcomes. This is a primarily democratic exercise, not a scientific one.

The end goal, or third stage, of processes of scrutiny and challenge is a decision about what can be reliably accepted as the basis of policy. Most models of the relationship between science and politics assume that nonexperts should set and revise ends in light of or response to scientific findings provided by experts. My analysis here suggests that there should be a prior step, where policy makers and the public form a judgment about the science itself. This judgment is not an appraisal of the scientific accuracy or merits of a finding, though in some cases political debate may well end up raising scientific doubts, but instead a judgment about what can be reliably assumed or accepted about the usability of science and its fitness for democratic purposes, given the limitations that have been exposed and examined. Chapter 3 discusses the role of scientific advisory committees in shaping this judgment.

The word "scrutiny" connotes careful inspection and surveillance. The processes of scrutiny that I have described, with their emphasis on criticism and contestation, have an adversarial stance built into them. One objection I anticipate is that the problems that I have highlighted about scientific claims and their influence in politics do not necessitate such democratic scrutiny of science but simply more transparency from scientists about the aims and priorities that have shaped their findings, with particular attention to uncertainty, incompleteness, and the assumptions they have made during the research process. If scientists communicate the strengths and weaknesses of their findings clearly and sincerely to the public, there would be no need for the further scrutiny of science by nonscientists.[52]

While more transparency and communication from scientists would undoubtedly improve the use of scientific findings and facilitate public deliberation on their basis, it would not be enough to address the issues that I have raised for several reasons. First, it can be difficult for scientists to recognize how their own values shape their research, and assess which of their assumptions are contentious, and which are obvious and unproblematic. This is especially true if their assumptions are shared within the scientific community and dictated by what is considered best practice. The AIDS clinical trials case illustrates this. Scientists at the time believed that good scientific methodology

---

52. For an argument along these lines, see Gregor Betz, "In Defence of the Value Free Ideal," *European Journal for Philosophy of Science* 3 (2013): 207–20.

required running homogeneous trials and did not recognize how their beliefs incorporated assumptions about the right practical approach to saving lives. The efforts of activists were crucial in making scientists accept that the fastidious and pragmatic approaches could both be valid, and ought to be regarded as appropriate for different purposes.

Second, it is hard for scientists to offer accounts of their omissions. An omission usually does not involve reflection or even awareness. Scientists pursue the explanations that they find plausible and significant. They may not have thought of some alternatives at all or dismissed them without much thought. Nonexperts usually have different perspectives, from which scientific omissions may be particularly visible. This is especially likely if an issue has significant practical consequences. The outsider standpoint can be instrumental in making scientists recognize what they have left out, and reconsider the way they frame and conduct their research. More diversity of perspectives within the scientific community could rectify these issues to some extent, but even a demographically diverse community of scientists may fail to generate the right kind of diversity in perspectives. Scientists as a professional group are not trained to deal with social and political issues. The critical abilities required for spotting omissions that impact different social groups, for example, are different than the training required for scientific research. The GMO case illustrates this difficulty. Most scientific studies focused on molecular-level changes that could be studied in lab settings, and neglected the concerns of local communities about biodiversity, sustainability, and food security.

A third point concerns the motivations of scientists. I mentioned in chapter 1 that this book brackets issues of corruption and fraud in science and the outright manipulation of findings to attain strategic ends. But even where there are no inducements for deception, it is reasonable to assume that scientists will be inclined to defend and promote their own findings, methods, and disciplinary approaches. It would be naive to expect scientists themselves to offer the most reliable account of the shortcomings of their own work. This is why research is peer reviewed. Yet conceptual, methodological, and evidentiary choices that follow from judgments about significance will not be detected through peer review because they do not necessarily affect the relationship between evidence and hypothesis. Since we cannot simply depend on scientists themselves to give us the most reliable account of how their discipline or approach ignores certain problems, people, or perspectives, we need external forms of scrutiny. This does not reflect a mistrust of scientists but instead an attitude of reasonable skepticism that rests on the belief that no individual or group should be assumed to be the most objective judge of the merits of their own work. Other scientists, especially those in other fields, can of course contribute to the process of public scrutiny.

Finally, scientists might disagree about the limits and uncertainty of findings, presuppositions of their discipline and methods, priorities driving research, and adequacy of findings for practical purposes. I have emphasized that the scientific and practical aspects of research design and disciplinary norms can be difficult to disentangle. It might be hard to identify the values and assumptions that have been used, whether they are subjective or dictated by scientific best practice, and whether choices are justified. In the face of disagreement between scientists, nonexperts must ultimately make a choice about what to accept. This requires engaging with the content of the disagreement. Disagreement itself can be valuable information for nonexperts, as I will argue in the next chapter, but it also creates more reason for proper scrutiny.

My argument so far has two important implications that are worth stating clearly. The first is that the ultimate judgment over the acceptance and use of scientific expertise must always lie with a political authority. It can be exercised by political representatives on behalf of citizens, or shared to some extent with ordinary citizens through institutions that allow for direct participation and decision-making. The primacy of political authority vis-à-vis the scientific is not justified by a claim to epistemic superiority on the part of democratic authority but rather on the grounds that the uncertainty, incompleteness, and context dependence of the available knowledge makes judgments about significance, priorities, and potential mistakes just as important as accuracy. Once we recognize how expertise is shaped by value judgments, it becomes clear that what are ordinarily thought to be matters of purely technical knowledge in fact require practical judgments that crucially depend on values. If we hold the belief that these are properly made by those authorized to make value judgments on behalf of citizens, then it follows that what is accepted as true for policy purposes should not simply be taken on authority from experts and must instead be determined democratically.

How radical is this claim? On the one hand, the philosophical critique that grounds this position is narrower than social constructionist views that maintain that the acceptance of scientific claims can be explained entirely through social relationships, hierarchies, and power dynamics in the scientific community. I do not challenge the empirical reliability of scientific findings, except to emphasize that they are necessarily uncertain and incomplete. I focus instead on how judgments of significance permeate the existing body of knowledge. On the other hand, the political conclusion I derive from this— namely, that the conceptual, methodological, and theoretical choices of scientists must be open to democratic scrutiny when science is used as the basis of political decisions—is neither obvious nor commonplace.

The second, perhaps more radical implication of my argument is that we might discover that much available knowledge is in fact unfit to help further democratic goals. This should not be surprising since most scientific research is not produced with political goals in mind. Even when it is, the aims and priorities assumed by scientists might not be ones that nonscientists would recognize as their own. When this is the case, policies may have to be made on the basis of less scientific knowledge rather than relying on scientific knowledge whose fitness for use is questionable. In the absence of scientific knowledge of the right sort, this may be the best we can do. The alternative—relying on knowledge provided by scientists without properly examining its suitability— might prevent the democratic determination of ends and produce bad outcomes. This also raises the possibility that there might be certain kinds of inquiry that should not be undertaken at all because the findings are likely to be misused in a practical context and thwart democratic ends. I develop this more controversial idea in chapter 6, where I argue that democracies should be more deliberate about what kinds of knowledge they want and consider how a democratic society can negotiate the limits of freedom of inquiry.

Democratizing Expertise

Insofar as science facilitates effective political action and helps expand the frontier of political possibilities, it does so in the ways that scientists imagine are useful for doing so. This is true not only in the selection of research questions, although that is most obvious, but also in conceptual, methodological, and theoretical choices. To understand how scientific assumptions and purposes shape findings and test their adequacy for the attainment of collective ends, I maintain that scientific expertise must be submitted to critical democratic scrutiny. This requires direct interaction between experts and nonexperts, a greater understanding of science by the public, and institutional spaces that facilitate the exercise of political judgment in matters of expertise, rather than separating the spaces for technical and political deliberation by design.

My conclusion is consistent with a rapidly growing literature that argues for the need for democratic participation on issues requiring expertise, but

I develop a novel justification for this position. Most arguments for democratizing expertise emphasize the need to guide science toward ends chosen by the public and the need to legitimate scientific authority democratically, but without paying sufficient attention to how the conduct of scientific inquiry affects the possibility of achieving these results.[53] I suggest that meaningfully democratizing expertise requires paying close attention to earlier stages of scientific inquiry. I started from an account of how the judgments about values and purposes made at the research stage shape scientific findings and the actions that can be pursued on their basis, and traced the implications of this for the democratic use of science.

There is always a gap between the observation that a scientific claim enjoys a certain degree of evidentiary support and the judgment that it is reliable to act on it for a particular purpose.[54] The latter is crucial for political action, and cannot be determined by science and scientists. Nor should it be. The question is about what it takes to move from scientific claims to the democratic judgments about their reliability for action. This chapter has begun to answer this question by exploring potential sources of failures in this move, and arguing that we must structure political processes to detect and respond to these failures. My aim was to be clearer, more precise, and philosophically rigorous about the implications of the status of scientific claims for their use in politics, and then think about what this implies about the proper attitude toward science in democratic decision-making. The next chapter moves away from scientific research and turns attention to the institutional site where science is translated for use in policy: the scientific advisory committee.

---

53. The notable exception is Kitcher, *Science, Truth, and Democracy*.
54. Cartwright and Hardie, *Evidence-Based Policy*.

3

The Paradox of Scientific Advice

IN 1801, THE FRENCH GOVERNMENT asked the *Ponts-et-Chaussées* assembly—the main corps of engineers responsible for the construction of roads and bridges—to look over a few proposals for the Saint-Quentin Canal.[1] After two months of examination, the assembly was divided between two proposals. The first project, by the military engineer Devic, proposed to drive two tunnels separated by an open-air canal in the middle to take advantage of a small valley in the plateau. It was perceived to be the safer and less risky option. The second, by the entrepreneur P. J. Laurent, aimed to build a single tunnel through the entire plateau. This project was more ambitious, and had both strong admirers and critics. It had been attempted in the 1770s, but then abandoned for political and economic reasons.

The Saint-Quentin Canal case was significant because it marked the first time in the eighty-five-year history of the assembly that a vote was used to settle a technical disagreement. During the old regime, the corps had made decisions hierarchically; inspectors would listen to the reports of engineers and then reach a decision. The revolution instituted the vote as the decision-making procedure for the assembly and gave voting rights not just to inspectors but also to chief engineers. Still, until the Saint-Quentin Canal, the assembly had resisted using a vote and made all its decisions through consensus. The disagreement between the proponents of the Devic and Laurent projects ran so deep, however, that they had no choice but to resort to a vote.

Despite the disagreement over the projects, members of the assembly were united by their aversion to the vote. To a body of technical experts, voting was

---

1. The following account is condensed from Frédéric Graber, "Obvious Decisions: Decision-making among French Ponts-et-Chaussées Engineers around 1800," *Social Studies of Science* 37, no. 6 (2007): 935–60.

an irrational method for reaching a decision. After all, the disagreement was over technical facts: the permeability and resistance of subsoils in the region, quantity of water required for navigation, dimensions of the tunnels, safety and practicality of navigation (profitability, the size of boats, and the organization of traffic), and so on. The engineers believed that division of opinion on a matter of fact meant that some people were either in error (and these could well be in the majority) or trying to further private interests rather than discovering the truth. They were also uneasy about revealing their uncertainty and unresolved disagreements to an increasingly hostile public and an intrusive administration. They wanted to make it clear that the assembly "made the right decisions, that it was the only institution capable of making them, and that no other decisions were possible."[2] The vote thwarted this ambition.

The *Ponts-et-Chaussées* engineers' antipathy toward the vote reveals an instinctive appreciation of an enduring fact: the authority and credibility of expert committees depends on their being perceived as making "obvious" decisions that rest on neutral scientific findings. At the same time, as the engineers were forced to recognize in the Saint-Quentin Canal case, experts often have to make decisions under conditions of uncertainty, incomplete knowledge, and persistent disagreement. From uncertain and disputed scientific claims about what is or may be true, they are expected to conjure reliable advice about what ought to be—which canal project should be selected or which policies will be effective. The purpose of this chapter is to analyze the inherently contradictory dynamics of decision-making in committees that provide scientific advice to the government, and suggest ways to resolve these tensions by rethinking the structure of the relationship between scientific bodies and political authority.

What I call the "paradox of scientific advice" consists in fact that the expectation that committees keep to neutral technical facts is not fully compatible with their fundamental task of providing useful advice to inform policy under conditions of uncertainty and incomplete knowledge. To be useful, advice must be scientifically sound, first and foremost, but it must also be designed to help decision makers set and attain their goals. This means that advice must be relevant, compatible with users' values and purposes (in ways that I will specify), simplified in the right way, and provided in a timely manner. Judgments about ends and values are thus frequently necessary for giving useful and timely scientific advice, as are subjective practical judgments in the face of uncertainty and disagreement. It is not only difficult to keep value judgments out of deliberations about which scientific claims to recommend as reliable for the purposes of a

2. Graber, "Obvious Decisions," 950.

policy but also the most plausible ways of trying to do so—continuing to seek evidence and deliberate over technical facts—render the advice less useful for its intended recipient, sometimes positively preventing the pursuit of ends by substituting irrelevant scientific concerns for the important practical ones.

This puts experts in a double bind: if they try to be more useful, they compromise the neutrality that is the source of their authority and legitimacy; if they find ways to remain as neutral as possible, they sacrifice usefulness and might inadvertently prevent productive courses of action. No matter what they do, they can end up being blamed—either for unhelpfulness or activism, and sometimes both at once. This is a difficult, if not impossible, charge for expert committees and renders the democratic role of scientific advice fundamentally unstable.

Identifying this paradox helps us better understand some of the dilemmas around expertise in politics and why they are so difficult to resolve. Late twentieth-century debates over expert advice have been in terms of the possibility or impossibility of neutral or value-free expertise and the challenges of drawing the boundary between science and politics.[3] By introducing the additional dimension of the relationship between the political usefulness of advice and value judgments that go into its production, I intend to complicate the aspiration to neutrality and rethink the proper role of advice in the institutional landscape of democracy. Thinking of the problem through this framework can also make us view recent proposals for the proper role of expert advice in politics in a different light—as attempts to resolve this problem by favoring different horns of the dilemma. Once we recognize this, we will be in a better position to evaluate the merits of existing proposals, and offer alternative remedies.

Institutions of Scientific Advice

The *Ponts-et-Chaussées* assembly was a precursor of the independent expert bodies that came to play an expansive advisory role in public policy in the second half of the twentieth century.[4] The unprecedented relationship between science and politics in the Napoleonic period was surpassed only in the

3. See, for example, Helen Longino, *Science as Social Knowledge: Values and Objectivity in Scientific Inquiry* (Princeton, NJ: Princeton University Press, 1990); Sheila Jasanoff, ed., *States of Knowledge: The Co-production of Science and Social Order* (London: Routledge, 2004); Heather Douglas, *Science, Policy, and the Value-Free Ideal* (Pittsburgh: University of Pittsburgh Press, 2009).

4. Historian Pierre Rosanvallon argues that these bodies came to represent a new form of legitimacy since the 1980s. See Pierre Rosanvallon, *Democratic Legitimacy: Impartiality, Reflexivity, Proximity*, trans. Arthur Goldhammer (Princeton, NJ: Princeton University Press, 2011).

postwar United States.[5] The provision of large amounts of public funding for science after World War II increased the scope and power of scientific research, and cemented the mutual dependence of scientists and the state. The size, complexity, and institutionalization of expert bodies offering scientific advice to the government grew rapidly in this period, in tandem with transformations in the scale and organization of science itself. Scientific organizations such as the National Research Council and the National Academy of Sciences, which had been founded earlier as purely scientific societies, began to take on more explicit and formal responsibility for providing regular advice to the government.[6] Scientists advising the government thus became a permanent feature of the institutional landscape of both democracy and science.

Today, the National Research Council produces between 200 and 250 scientific assessment reports each year, on issues ranging from new technologies and environmental impacts to the public understanding of science.[7] A typical assessment involves the participation of dozens and sometimes hundreds of scientists, with over six thousand scientists participating in assessment work each year.[8] Producing assessments for policy has become such a highly structured and complicated activity that it is now recognized as a distinct form of scientific work.

Scientific advisory bodies have an unusual status in the institutional landscape of politics because they are composed of independent scientists rather than elected politicians or appointed bureaucrats. Their members are selected for distinguished achievement in their professional fields, which allows them to combine high-quality expertise with a plausible degree of detachment from politics. These two features are the source of the authority and credibility of these bodies, but also complicate their democratic status. Independent expert bodies represent a rival source of authority in a democracy that can at once

5. Charles C. Gillispie, *Science and Polity in France: The Revolutionary and Napoleonic Years* (Princeton, NJ: Princeton University Press, 2004). For an account of the setting up of this contemporary order, see Peter Weingart, "Scientific Expertise and Political Accountability: Paradoxes of Science in Politics," *Science and Public Policy* 26, no. 3 (1999): 151–61.

6. Michael Oppenheimer, Naomi Oreskes, Dale Jamieson, Keynyn Brysse, Jessica O'Reilly, Matthew Shindell, and Milena Wazeck, *Discerning Experts: The Practices of Scientific Assessment for Environmental Policy* (Chicago: University of Chicago Press, 2019).

7. US National Academy of Sciences, "About Our Expert Consensus Reports," http://dels .nas.edu/global/Consensus-Report.

8. US National Academy of Sciences, "What We Do," http://www.nationalacademies.org /about/whatwedo/index.html.

strengthen and threaten democratic rule. On the one hand, reliance on the knowledge and competence of expert bodies can improve democratic outcomes and enhance citizens' welfare, thus playing a key role in the success of democratic governments. On the other hand, the superior knowledge of experts can be used to justify leaving more and more decisions up to them, which risks diminishing the scope for democratic decision-making. The ready availability of expertise in the policy realm can become an excuse for defining more and more policy areas as requiring complex technical knowledge, thereby creating a self-sustaining system of expert-led policy making.

Current institutional arrangements respond to this familiar dilemma of expertise by reverting back to the traditional Weberian division of labor that I criticized in the last chapter: experts are meant to handle the facts, based on an objective analysis of the evidence, while citizens and their representatives decide on the ends to pursue based on their values and preferences. This model offers a way to reconcile the scientific authority of expert bodies with political authority. While the problem of expertise consists in the worry that the authority of experts will become the source of political authority, this model preserves an autonomous realm for the latter by stipulating that the two kinds of authority rule over different domains. Experts may command deference over beliefs, but the democratic will reigns supreme over actions. Scientific authority is therefore given a subordinate role in political decision-making; it is treated as an instrument for the attainment of political ends. In practice, experts might claim political authority by making pronouncements on what they think should be done, and their advice may be taken as authoritative because they are recognized as authorities. This, however, would constitute an unacceptable sliding of scientific authority into political.

Despite the practical challenges of drawing the boundaries between science and politics, this model still underlies both the formal mandate of scientific advisory committees and self-understanding of the scientists who serve on them. The National Academy of Sciences describes its mission as providing an objective assessment of the latest scientific evidence that the government may need before it makes policy decisions.[9] Scientific advice is intended to precede and inform decision-making by the government without making prescriptions. The Intergovernmental Panel on Climate Change, the most well-known and visible scientific advisory body of the past three decades, has

---

9. US National Academy of Sciences, "Mission," http://www.nasonline.org/about-nas /mission/.

likewise embodied this division of labor logic.[10] Its self-stated aim is to assess the scientific literature relevant to understanding climate change in a way that is "policy-relevant and yet policy-neutral, never policy-prescriptive."[11] This summarizes the approach of most current scientific advisory bodies and gives the clearest expression to the underlying ideal of neutrality. The division of labor logic is evident in the sequential structure of advice and policy.

The ideal of neutrality prevails in the culture of the scientific community too. Geoscientist Michael Oppenheimer and colleagues show that many scientists who participated in the scientific assessments for acid rain, ozone depletion, and sea-level rise believed it was crucial for them to be seen as neutral in order for the assessment to be effective.[12] Those who had publicly expressed their policy recommendations were not invited to participate in the assessment lest they make the committee appear biased, even if these scientists were the most competent researchers working in the relevant area. Most scientists interviewed reported that they believed that reliably informing policy while remaining neutral was possible, and indeed necessary and desirable. Some scientists saw neutrality to be crucial for the public credibility of science, especially given the declining levels of trust in science. Others gave a democratic justification of the division of labor, arguing that making political judgments is neither the right nor responsibility of scientists, and that their private opinions are irrelevant to their public responsibility for informing policy.[13]

I define neutrality as a stance that requires scientists to refrain from making judgments about moral and political values, limiting their claims to what is, rather than what ought to be. The aim of neutrality is to ensure that scientific advice can serve different value outlooks evenhandedly and does not privilege some over others.[14] Neutrality in this sense does not suggest that science itself is or should be value free, but requires advisers to adopt an attitude of restraint

---

10. Matthew J. Brown and Joyce C. Havstad, "The Disconnect Problem, Scientific Authority, and Climate Policy," *Perspectives on Science* 25, no. 1 (2017): 67–94.

11. Intergovernmental Panel on Climate Change, "Organization," https://archive.ipcc.ch /organization/organization.shtml.

12. Oppenheimer et al., *Discerning Experts*, 184–87.

13. For accounts of different historical defenses and uses of the neutrality ideal, see Robert N. Proctor, *Value-Free Science? Purity and Power in Modern Knowledge* (Cambridge, MA: Harvard University Press, 1991); Liam K. Bright, "Du Bois' Democratic Defence of the Value Free Ideal," *Synthese* 195, no. 5 (2018): 2227–45; Andrew Jewett, *Science under Fire: Challenges to Scientific Authority in Modern America* (Cambridge, MA: Harvard University Press, 2020).

14. Hugh Lacey, "Rehabilitating Neutrality," *Philosophical Studies* 163, no. 1 (2013): 77–83.

and leave aside moral and political judgments during the advisory process. Scientists can be more or less neutral, even if absolute neutrality is not attainable. Neutrality is different than objectivity, which I take to refer to the empirical reliability of scientific claims. While some conceptions of objectivity may require neutrality, others do not. This understanding of neutrality should be distinguished from two nearby alternatives. The first is neutrality as the active balancing of different values and interests with the aim of treating them all equally. Douglas calls this approach "reflectively centrist."[15] The problem with this is that centrism is itself a moral and political stance that must be justified. Some values may be objectionable, and balancing even unobjectionable values may be worse than selecting some over others. There is no reason to think balancing is desirable as a rule. The second alternative is to define neutrality as the position that emerges from critical interaction and negotiation among different values. I classify this later as one of the useful stances that a scientific advisory committee can take, but it would be conceptual stretching to call it neutrality, unless we treat neutrality entirely as a constructed pose.

My aim in this chapter is to take a more critical look at the stability and desirability of the aspiration to neutrality by examining the mechanics of decision-making in advisory committees. I will focus especially on the interplay of evidentiary and practical considerations in the provision of science advice. Note that I will not be offering an empirical account of how well scientific committees live up to the charge of neutrality, or respond to the tension between neutrality and usefulness. Studies in the sociology of science have shown that the claim that scientists or scientific advisory bodies can be fully neutral or value free is usually false in practice. Sheila Jasanoff demonstrates that science and politics are "co-produced" in advisory contexts, and Thomas Gieryn argues that the boundary separating science from politics is actively constructed, negotiated, and defended by scientists and politicians working at the intersection of these two spheres.[16] These sociological arguments, however, evaluate neutrality as a constructed pose rather than as an ideal that can be approximated more or less well; they are therefore normatively inert.[17] It

15. Heather Douglas, "The Irreducible Complexity of Objectivity," *Synthese* 138, no. 3 (2004): 453–73.

16. Jasanoff, *The Fifth Branch*; Jasanoff, *States of Knowledge*; Gieryn, "Boundary-Work and the Demarcation of Science from Non-Science."

17. For more on the theoretical distinction between ideals as descriptive and normative models, see Charles Mills, "'Ideal Theory' as Ideology," *Hypatia* 20, no. 3 (2005): 165–84.

might well be that neutrality should also be discarded as an ideal, but this conclusion requires more normative argumentation. In the following, my aim will be to examine neutrality as a regulative ideal for advisory bodies, and assess its desirability from a democratic and epistemic perspective.

Sufficiency and Simplification

The advisory process is different from research in important ways. First, advisory bodies usually do not undertake or commission new research; they evaluate existing peer-reviewed literature. Second, their advice is oriented explicitly toward practical goals, and intended to produce an action or decision. The need for advice arises from the identification of a problem that requires a response. The final product of an advisory committee must be a set of claims that decision makers can accept as true in their deliberations and planning toward solving the problem.[18] What constitutes appropriate scientific advice will therefore depend, in ways that I will specify, on the values and purposes of users. Scientific advice that is right for one person or in one context will not be so for another person or in another context. Advice is more useful—in the purely instrumental sense of helping the government set and attain democratic goals reliably—insofar as it can incorporate the values and purposes of citizens and decision makers.

One of the main ways in which useful advice requires practical judgments is in the determination that the available evidence is sufficient for an action or decision under uncertainty. The literature on inductive risk argues that scientific inference always requires a judgment about the sufficiency of the evidence for accepting a hypothesis.[19] Inductive inference is inherently open-ended; no amount of empirical observation can guarantee the truth of an inductive

18. For more on this attitude of "acceptance in a context" or "acceptance for a purpose" and its role in practical reason, see Michael E. Bratman, "Practical Reasoning and Acceptance in a Context," *Mind* 101, no. 401 (1992): 1–15. Stephen Turner's concept of "fact-surrogates" expresses a similar idea. See Stephen Turner, *Liberal Democracy 3.0: Civil Society in an Age of Experts* (London: Sage, 2003).

19. The original debate was initiated by C. West Churchman, "Statistics, Pragmatics, Induction," *Philosophy of Science* 15 (1948): 249–68; C. West Churchman, "Science and Decision Making," *Philosophy of Science* 23, no. 3 (July 1956): 247–49; Richard Rudner, "The Scientist qua Scientist Makes Value Judgments," *Philosophy of Science* 20, no. 1 (January 1953): 1–6. These arguments were revived in Douglas, *Science, Policy, and the Value-Free Ideal*. For responses to and extensions of Douglas's argument, see Kevin Elliott and Ted Richards, eds., *Exploring Inductive Risk: Case Studies of Values in Science* (New York: Oxford University Press, 2017).

generalization. Scientists must therefore judge whether the evidence is sufficient to accept a hypothesis. This judgment must be relative to an assumed purpose: Sufficient for what? This gap in inductive inference forces the scientist to consider the purposes for which the accepted hypothesis might be used and decide on the basis of the potential consequences of making a mistake. It is always possible to accept a wrong hypothesis or fail to accept a true one; this is the inherent risk of induction.

Considerations about inductive risk also arise in the decision-making of advisory committees. An advisory committee cannot be oblivious to the consequences of its decisions about the sufficiency of evidence.[20] One of its main tasks is to evaluate the strength of the available evidence with respect to possible real-world consequences. In fact, the charge of advisory committees is often expressed in terms of assessing the sufficiency of evidence on a particular question. An advisory committee asked to assess whether certain substances are carcinogenic or the fluoridation of the water supply is safe is essentially asked to decide whether the evidence can be considered sufficient to reach these conclusions. This requires considering the potential consequences of these judgments, and making normative judgments about the relative badness of different types of mistakes, such as false positives and false negatives.

These judgments depend on the assumed purposes and perspectives. Agents with different interests will demand different levels of evidence in order to accept a scientific claim to be sufficiently reliable to act on. Any choice of evidentiary threshold thus implicitly weighs in favor of one set of interests or purposes over another. Advisory committees might attempt to circumvent this problem by using widely shared scientific standards, but in doing so they import standards developed for the particular purposes of scientific inquiry to areas where they might not be appropriate. For instance, the common scientific practice of looking for 95 percent statistical significance may be appropriate for the scientific preference for avoiding mistakes and erring on the side of caution, but would be problematic on issues that require quick action and where mistakes would not be too costly.

Another important function of an expert committee is to simplify complex information for decision makers and the public. This involves both the

---

20. Katie Steele and Stephen John similarly point out that the inductive risk debate is productively understood as a problem for policy advising. See Katie Steele, "The Scientist qua Policy Advisor Makes Value Judgments," *Philosophy of Science* 79, no. 5 (2012): 893–904; Stephen John, "Inductive Risk and the Contexts of Communication," *Synthese* 192, no. 1 (2015): 79–96.

translation of technical language into language accessible to laypeople as well as a reduction of complexity and volume by discarding some findings and eliminating details and nuances.[21] Scientific committees are frequently the first and sometimes only ones to take on the responsibility of simplifying scientific knowledge for others. This simplification serves two crucial democratic functions. It allows policy makers to make decisions on the basis of technical information, which they might otherwise be unable to understand or process, and opens up expert judgment to democratic scrutiny. Most citizens cannot evaluate expertise unless it is sufficiently and reliably simplified. Of course, such scrutiny will be possible only if scientists doing the simplification avoid concealing uncertainties and gaps in a way that precludes debate over the content and implications of the science.

I want to make two points about the relationship between simplification and neutrality—one obvious, and one not so obvious. The obvious point is that there is no neutral way of aggregating, summarizing, and simplifying information. Any simplification process involves choosing what to include and what to leave out. This choice is usually made with reference to what is considered significant and relevant, which necessarily requires purpose- and agent-relative judgments. Insofar as an advisory committee ignores considerations about what will be politically significant and relevant, its advice will be less useful. But any attempt to judge significance and relevance will force advisers to adopt a particular perspective, which will move them away from neutrality. Both scientists and politicians understand this, which is why executive summaries are often the most contentious parts of an assessment report. To give one example, the Intergovernmental Panel for Climate Change faced pressure over the wording of its summary for decision makers, and eventually had to change it in response to opposition from the Saudi and Kuwaiti delegates.[22]

The less obvious point is that the most useful summary for a decision maker will not necessarily consist of claims that are evidentially best supported, especially under conditions of uncertainty and complexity. Accuracy is usually instrumental to the realization of one's purposes, but it can be counterproductive in some cases and insufficient in others. This is because accuracy can stand in

21. For more on the translation of science into the ordinary language of practice, see Jürgen Habermas, *Toward a Rational Society: Student Protest, Science, and Politics*, trans. Jeremy Shapiro (Cambridge, UK: Polity Press, 1987).

22. Alfred Moore, *Critical Elitism: Deliberation, Democracy, and the Problem of Expertise* (Cambridge: Cambridge University Press, 2017), 137–38.

tension with other features of scientific findings that are important for the attainment of practical goals. Since scientific advice is oriented toward action rather than just truth, it will be improved by attention to these practical considerations as opposed to merely the quality and strength of the evidence. In some cases, there will be a trade-off between simplicity and accuracy. Scientists often use scientific theories that they know to be false because these theories are easier to use than more accurate but complex ones. Engineers rely on the Newtonian theory of gravitation for most purposes, even though it is, strictly speaking, false. A similar trade-off between simplicity and accuracy arises in expressions of uncertainty. The translation of uncertainty for policy makers requires moving from fine-grained probabilistic measures that more accurately represent scientists' beliefs to cruder ones that are easier to communicate and use.[23]

In some cases, committees agree on the relative merits of competing findings, but must decide how to trade different values such as accuracy and usability. In other instances, there will be genuine uncertainty and disagreement among committee members about the relative merits of findings. They will nonetheless have to select a small number of claims to present to policy makers. One way to simplify deliberations in these cases is to focus on the evidence alone, and try to reach an agreement on the theories or models that are scientifically best supported. The *Ponts-et-Chaussées* engineers responded to their deep disagreement by adopting this strategy. This is not always a good policy for helping decision makers, however, because trading off accuracy against other values may in fact increase the chances of attaining particular goals. There may be properties of a scientific theory or model that will make it more likely that decision makers will attain outcomes that they want to bring about if they accept (or reject) a theory or model with those properties, even if turns out not to be true. The clearest examples are what Jacob Ross calls "deflationary" theories, which predict that all options facing an agent will yield a payoff of the same value; this means that it doesn't matter what the agent chooses.[24] In situations where scientists have some credence in theories that are deflationary and those that are not, it would be rational to discard the deflationary ones, even if they are evidentially better supported.

Here's an illustration. Suppose there are several local climate models. Some suggest that there will be severe droughts in East Africa and many people will

---

23. Steele, "The Scientist qua Policy Advisor Makes Value Judgments."

24. Jacob Ross, "Rejecting Ethical Deflationism," *Ethics* 116, no. 4 (2006): 742–68.

die. They also indicate that no intervention can prevent this outcome. The others make similar predictions about the fatality of the drought, but suggest that changing irrigation methods and introducing new crop varieties at a small cost can mitigate its severity and save many lives.[25] If scientists think that these models have roughly similar plausibility, it is worth discarding the deflationary ones in their advisory summary. But they can make this judgment only if they can make some assumptions about the ends that a policy maker might wish to pursue. Ross also demonstrates that acting on theories that predict fewer good options and theories with a higher spread of outcomes can increase the chances of maximizing good outcomes and minimizing the cost of mistakes, even if these theories enjoy less evidentiary support than alternatives that predict more good outcomes and have lower spread. Scientists could provide more useful advice by selecting theories with these properties, but only if their assumptions about what constitutes a good outcome versus a mistake align with those of users.

Users' risk aversion should also play a role in the selection of the models that it would be best for them to act on. A risk-averse community would have a better chance of attaining its ends by acting on models that give greater weight to bad outcomes, whereas a risk-loving one would be rational to choose more optimistic models. This would hold even if an alternative model enjoyed more evidentiary support. Just how much more evidence one needs to have before accuracy trumps all other considerations about risk and payoffs depends in part on risk aversion and the values that users assign to different outcomes. These show how an advisory committee's decision about which results to report could be improved by considering the ends toward which the knowledge would be used as well as the values and preferences of users. Scientists cannot provide useful advice without making some assumptions on these points, given the impossibility of presenting a summary that is both comprehensive and neutral.

Advisory committees weigh the scientific evidence along with a purpose-relative normative or practical consideration, such as its significance, relevance, sufficiency, goodness, badness, riskiness, or usability. Scientific findings must be evaluated according to these criteria in order to convert science into good advice. Philosophical debates about the role of values in science usually focus on whether value-free science is possible. I have argued that value-free

25. This example was inspired by Esther Ngumbi, "How to Tackle Repetitive Droughts in the Horn of Africa," *Al Jazeera*, February 14, 2017, https://www.aljazeera.com/indepth/opinion /2017/02/tackle-repetitive-droughts-horn-africa-170214090108648.html.

scientific advice is not only not possible but also that trying to free advice from values makes it significantly incomplete, and thus less useful than advice that explicitly considers the ends and values of a decision maker. The fact that these practical judgments depend crucially on an understanding of the strength and uncertainty of the evidence creates a prima facie epistemic (though not yet moral or political) case for scientific committees to make these judgments on the grounds that they understand the evidence best. This underscores the trade-off between the neutrality of a committee, which could be fulfilled by refraining from making these normative judgments, and the usefulness of its advice, which could be enhanced by making them.

Aiming for Neutrality

There are two ways of addressing the tensions identified. The first is for scientists to try approximating the ideal of neutrality; the second is to abandon the ideal in various ways. Both approaches might concede that neutrality is impossible to achieve fully, but the first sees it as a valuable regulative ideal whose close approximation is possible and desirable, whereas the second views approximation as undesirable and maintains that scientific committees must make the relevant normative judgments as responsibly as they can. These approaches can be seen as corresponding to political scientist Roger Pielke Jr.'s classification of advisory styles. What he calls "pure scientists" and "science arbiters" are examples of neutral advisers who simply provide information. His "honest brokers" and "issue advocates," by contrast, favor usefulness and engage closely with the values and choices of the decision maker.[26] Let me take each view in turn.

One way scientific advisers can approximate neutrality is by keeping the discussion to evidentiary matters wherever possible, and using scientific or truth-related values and purposes, as opposed to ethical and political ones when values and purposes are unavoidable. Gregor Betz argues that scientific advisory bodies should avoid making value-laden judgments about the sufficiency of evidence by carefully reporting the uncertainty of different hypotheses.[27]

26. One important difference between my argument and Pielke's is that he closely associates these stances with the provision of one or a few policy recommendations. In my argument, by contrast, usefulness is not defined with respect to policy recommendations but advisers' direct engagement with the values of users during deliberations. See Pielke, *The Honest Broker.*

27. Gregor Betz, "In Defence of the Value Free Ideal," *European Journal for Philosophy of Science* 3 (2013): 207–20.

Where uncertainty forces a choice between different types of error, scientists should weaken their language so that the available evidence confirms their conclusions beyond a reasonable doubt. This would minimize the risk of error and avoid moral judgments about the relative desirability of different types of error. The result of these efforts would not be perfect neutrality, but it could plausibly be described as an approximation. Indeed, there is evidence that scientists often respond to the charge of neutrality precisely in this way. The *Ponts-et-Chaussées* engineers' technical debates present a standard example of an expert committee delving deeper into technicalities in order to avoid making practical judgments. Sociologists Boris Hauray and Philippe Urfalino diagnose a similar phenomenon in the European Union's medical licensing committees. They show that scientific advisers to the European Union responded to the increasing pressure to remain neutral toward competing national interests by making decisions based on scientific arguments alone.[28] Neutrality across countries was achieved at the expense of nation-specific socioeconomic priorities.

There are several problems, though, with this approach to approximating neutrality. The first is that it can end up masking the ways in which advice is in fact nonneutral. On the most cynical view, scientists could envision the downstream political implications of different scientific claims and tailor scientific arguments to advance their political preferences. This is one of the standard fears about expert committees.[29] Second, even if we set aside abuses of neutrality, and focus on sincere attempts by scientific advisory committees to remain neutral by keeping to scientific values and language as much as possible, this would be undesirable because it renders the advice less useful. Advice that focuses on resolving purely scientific issues and reducing uncertainty without addressing practical concerns that matter to decision makers and the public might simply

28. Boris Hauray and Philippe Urfalino, "Mutual Transformation and the Development of European Policy Spaces: The Case of Medicines Licensing," *Journal of European Public Policy* 16, no. 3 (2009): 431–49.

29. An alternative defense of the neutrality ideal by Harry Collins and Robert Evans recognizes that scientists may end up smuggling in their value judgments, and offers the solution of a committee of social science experts—"owls"—tasked with assessing the strength and substance of the scientific consensus impartially. This solution, however, cannot avoid the neutrality-usefulness dilemma since the owls would likewise face the challenges of determining the sufficiency of evidence, reporting a simplified summary, selecting relevant models, and so on. It also introduces an additional layer of uncertainty, disagreement, and possible value judgments, this time among the owls. See Harry Collins and Robert Evans, *Why Democracies Need Science* (Cambridge, UK: Polity Press, 2017).

be ignored. In the process, scientists might waste precious time debating complex but digressive technicalities. The *Ponts-et-Chaussées* engineers' prolonged discussions exemplify this danger. Additionally, valuable information might be lost in the process. If scientists apply standards of evidence that are appropriate for scientific purposes and report information based on its scientific merits alone, they will fail to give due weight to findings that might have important practical implications yet whose scientific merits are more tenuous.

There is a paradoxical trade-off here: introducing nonevidentiary considerations into advisory committee deliberations can increase the chances of scientific error, even while it increases the chances of attaining the desired practical goals. Thinking decision-theoretically can help us see why. The best course of action for a person is determined by the values that they attach to different outcomes, along with the probability of bringing them about. If a person attaches a large positive or negative value to an outcome, this outcome will acquire substantial weight in their decision calculus, even if it has a small probability of occurring. Possible outcomes that scientists leave out of a report on the grounds that the evidence for them is not strong can be crucial for someone who attaches a sufficiently large positive or negative value to them. Similarly, it can be rational to bracket scientifically well-supported theories that make no difference to the outcome. These are the deflationary theories mentioned earlier.

This line of reasoning bears resemblance to Pascal's wager. Pascal argued that it would be rational to wager that God exists, even if the available evidence is small, given the infinite utility of salvation. If Pascal had asked a committee of scientists to advise him on the question of the existence of God, and it had failed to report the possibility that God exists on the grounds that there isn't sufficient evidence for the claim, Pascal would have been deprived of infinite expected utility (assuming that there is at least a slight possibility that God exists). In other words, a committee that focuses on purely scientific merit can end up dismissing possibilities that would be significant for decision makers given how they value different outcomes. This problem arises from the pressures of simplification. If scientists could simply turn over all the evidence and report the probabilities of all possible outcomes, as Betz has suggested they should, then policy makers would be well equipped to judge what they ought to do in light of democratic values and preferences. But such comprehensive reporting is practically impossible given the complexity and size of the universe of possible outcomes as well as unhelpful given the limits of decision makers' time, attention, and cognitive capacities.

Note that this argument does not require the controversial move of cultivating real faith or belief in possibilities that appear quite unlikely, as would be required in Pascal's wager. It is sufficient to treat them as if they are true—to "accept" them as true, in Michael Bratman's terminology—for the purposes of deliberation or planning, even while recognizing that they might turn out to be false.[30] I use infinite utilities and infinitesimal probabilities only to illustrate the logic of this argument; the core reasoning is robust against the objection that it would not be rational to act on such extremely unlikely possibilities or assign extremely high values to any outcome. The argument would apply equally well to cases that are less extreme, such as where scientists and decision makers agree that they should only consider scientific theories that are fairly plausible and not assign infinite utility to any outcomes. Advisory committees would still face a choice among several more or less well-supported theories that are all fairly plausible. If they made the choice only with reference to scientific considerations in order to maintain neutrality, information valuable from the decision maker's perspective would still be lost.

Conceptualizing the challenges of advice through this decision-theoretical lens can change our interpretation of some well-known public controversies around science by revealing the possibility that nonexperts who appear to be resistant to accepting scientific advice may in fact be attaching extremely high values to low-likelihood outcomes. While doing so may be unreasonable or unethical, this is a distinct issue from a failure to understand or accept scientific evidence. Recent problems with parents' refusal to vaccinate their children can be reinterpreted in this light. The probability that a vaccine will harm an individual child is usually so small that public health officials advising the government and public may discard the likelihood as negligible, and instead emphasize the safety of the vaccine. Parents who refuse to vaccinate their children are then portrayed as denying sound scientific advice. An alternative way of understanding the problem, however, is to recognize that these parents are acting on the small risk of injury that a vaccine always poses and assigning extremely large negative utility to the rare outcome of their child contracting a severe reaction to the vaccine. Since the benefits of vaccination accrue at the population level, this calculus may be rational for the individual parent.[31]

30. Bratman, "Practical Reasoning and Acceptance in a Context."

31. Anna Kirkland, *Vaccine Court: The Law and Politics of Injury* (New York: NYU Press, 2016). See also Gürol Irzık and Faik Kurtulmuş, "Well-Ordered Science and Public Trust in Science," *Synthese* (2018), doi.org/10.1007/s11229-018-02022-7.

What the refusal to vaccinate betrays, more than anything, is a failure of solidarity and care for the well-being of other members of society, particularly for the most vulnerable, who are most likely to suffer from a potential outbreak.

Seeing the case this way may not change our verdict on the rightness or wrongness of failing to vaccinate, but it crucially alters our understanding of the causes. This view implies that trying to change the values and preferences of parents would be a more effective strategy for changing behavior than claiming that they fail to understand the science. Additionally, it explains why advisory committees' purportedly neutral attention to the scientific merits of claims about vaccine safety without attention to the concerns and priorities of the public can go badly wrong. While the pressure for simplification and the balance of the scientific evidence may justify a committee's decision not to stress certain small possibilities, it can also lead to a loss of information that others would regard as important, which in a hostile environment, can turn to distrust of the motives of the committee.

I have argued that trying to approximate neutrality by keeping to the purely technical can be a bad thing; it can lead to irrelevance, waste of time, loss of valuable information, and even distrust of sincere efforts to remain neutral. Falling back on scientific standards is inappropriate for advice whose main purpose is practical. Science and politics are activities with different goals and purposes. Although they both share an interest in the scientific evidence, they have practical goals too, exerting different practical pressures. There is no reason to think that the standards and procedures conducive to success in scientific inquiry would also be the ones conducive to the attainment of political goals.

Useful Advice

The ideal of neutrality discussed so far requires scientists to adopt an attitude of detachment from political debates and conflicts. An expert committee that is neutral in this sense refrains from considering the interests and purposes that might be affected by its advice, let alone trying to shape them, even if this restraint compromises its relevance and usefulness. On this model, scientists are expected to stand outside politics in order to protect their credibility and prevent the encroachment of their authority into the domain of political decision-making. The price of keeping scientists out of politics this way is to force them to be unengaged and even indifferent to the political consequences of their work. I now want to examine an alternative ideal for scientific advice, which moves away from the aspiration to neutrality understood as detachment from moral and political values and offers a more overtly political role for expert committees.

On this alternative vision, advisory committees are expected to engage with the practical implications of their scientific advice and make necessary value judgments rather than falling back on reporting facts. To be clear, they are still not meant to advocate for specific policy decisions, but they are seen as appropriate sites for quasi-political deliberations, where considerations about ends and values should be brought up and discussed alongside scientific or technical matters. In practice, this means advisory committees trace the implications of scientific findings for different stakeholders, consider the moral and political significance of evidence alongside its strength, tailor their reports to the relevant political priorities, determine evidentiary standards in light of the consequences of different kinds of error, and offer policy options. Scholars who study advisory bodies empirically have noted that expert committees often engage in this kind of quasi-political work in practice, but this has rarely been spelled out and defended as an ideal; it is usually viewed as an inevitable tendency at the intersection of science and politics.[32] To theorize scientific committees as potential sites for a kind of political activity, we must clarify how advisory committees could fulfill the political aspects of their task, and how this can be legitimized in the landscape of democratic institutions that fall outside the legitimating power of electoral representation. This requires specifying how different values and purposes should be represented in committee discussions, and how nonexperts can evaluate the committee's conclusions.

The Ethical Scientist

One solution to these challenges is to maintain that scientists have a moral responsibility to consider the consequences of their findings and set evidentiary standards with reference to the social and moral consequences of different types of error.[33] This view focuses primarily on individual scientists, who

---

32. Jasanoff, *The Fifth Branch*; Susan Owens, *Knowledge, Policy, and Expertise: The UK Royal Commission on Environmental Pollution 1970–2011* (Oxford: Oxford University Press, 2015).

33. Douglas, *Science, Policy, and the Value-Free Ideal*. See also David Resnik, "Dual-Use Research and Inductive Risk," in *Exploring Inductive Risk: Case Studies of Values in Science*, ed. Kevin Elliott and Ted Richards (New York: Oxford University Press, 2017), 59–78; Joyce C. Havstad and Matthew J. Brown, "Inductive Risk, Deferred Decisions, and Climate Science Advising," in

bear the burden of making the necessary value judgments as their ethical responsibility. For instance, if a toxic air pollutant is associated with spikes in respiratory deaths but the evidence for causality remains uncertain, scientists should consider the moral consequences of emphasizing the dangers versus the uncertainty.[34] Since the consequences of failing to underscore real dangers would be worse than raising a false alarm, scientific advisers have a moral obligation to stress the dangers.

Yet the plausibility of this example rests on the obviousness of the moral case. In many situations, especially when the science is new and uncertain, the moral dilemmas and distributive problems involved will be largely speculative and subject to disagreement. The kinds of controversies that emerge around scientific issues are not matters where correct answers can be discovered through ethical deliberation. Instead, these issues require finding common ground, and reaching agreement about the values, purposes, and interests that must take priority when different individuals and groups have irreconcilable differences. In other words, they require political judgment rather than careful application of ethical reasoning. The exercise of such judgment requires representing the interests, values, and goals of different people through legitimate channels of representation. In practice, scientific advisers could try to discharge this responsibility by acting as informal political representatives, and channeling different stakeholders' interests and values instead of just consulting their own moral judgment.[35] But since scientists are not trained, qualified, or authorized to act as political representatives, this would have no more democratic legitimacy than if they relied on their personal ethical judgment alone.

This approach thus ignores the challenges of political representation as opposed to providing a solution to them. It allocates to scientists the power to make value judgments in the name of others, despite their lack of legitimacy as representatives. This is precisely the kind of scenario that the Weberian ideal was intended to prevent. Still, this model serves as a valuable foil for alternative proposals since it expresses a default position: unless we can specify how

Exploring Inductive Risk: Case Studies of Values in Science, ed. Kevin Elliott and Ted Richards (New York: Oxford University Press, 2017), 101–25; Anya Plutynski, "Safe or Sorry? Cancer Screening and Inductive Risk," in *Exploring Inductive Risk: Case Studies of Values in Science*, ed. Kevin Elliott and Ted Richards (New York: Oxford University Press, 2017), 149–70.

34. Douglas, *Science, Policy, and the Value-Free Ideal*.

35. On informal political representatives, see Wendy Salkin, "Informal Political Representation: Normative and Conceptual Foundations" (PhD diss., Harvard University, 2018).

political representation can be possible within a scientific committee, scientists will have nothing but their own moral judgments to fall back on, whether they choose to exercise it as representatives or reflective moral agents.

Scientists as Representatives

Brown has written most extensively about representation in contexts of expertise, so he offers a more direct response to these concerns. He argues that scientific advisory committees should be conceived as sites where social and political representation can be achieved through an active, careful balancing of perspectives. This means bringing together stakeholders with experts in hybrid committees and ensuring a diversity of views on each side. Brown maintains that stakeholders and experts should not be conceived as representing social and professional interests, respectively—that would replicate a problematic division of labor—but instead that both should be conceived as representing different social perspectives.[36] The goal is to strike a fine balance between ensuring that the composition of advisory committees mirrors social perspectives as closely possible, but also avoiding the extremes of purely partisan alignments and apolitical scientific representation.

One problem with this view is that it is difficult to specify in advance the perspectives that will be relevant in composing a committee. Should perspectives be understood in terms of demographics, geographic location, nationality, or professional commitments? Brown suggests that the answer will be given by the purpose of the committee, but it is unlikely that identifying a purpose will be sufficiently determinate, especially if the purpose is to provide advice on a new scientific development and its practical implications. This creates a chicken-and-egg problem: without proper representation on the committee, the resulting advice may be biased; without a clear and unbiased sense of the issue, advisory committees cannot be composed in the properly representative way. The problem of deciding on the purpose and membership is particularly acute for scientific advisory committees that work at the forefront of scientific research since the cleavages that will organize the subsequent political uptake of the issue are still unclear. Without a social consensus about the main perspectives on an issue, the appropriate composition of committees will be indeterminate, and any attempts to settle the issue will be open to the charge of arbitrariness.

36. Brown, *Science in Democracy*.

But there is a deeper problem: even the careful balancing of perspectives within a scientific committee cannot stand in for democratic representation. There are intrinsic limits to how representative a committee can be due to its small size and special composition of experts. Members of an expert committee are quite different than the rest of the public, and belonging to a demographic, geographic, or professional group does not mean one will be a representative member of that group. Moreover, it cannot be assumed that large groups will have monolithic views. Brown himself admits the impossibility of replicating the full diversity of society within a single committee, so he suggests that committee members should still deliberate on their own judgment rather than thinking of themselves as direct representatives of a group. In doing so, Brown reveals the limits of political representation through expert committees and falls back on something like the ethical scientist model, relying on competent reflection and high-quality deliberation by scientists.

In sum, recent accounts that reimagine the role of scientific advisory committees in a democracy do not offer fully satisfactory answers to the challenges of representation that arise once it is acknowledged that scientists must make a range of value judgments. Advisory committees must undertake some ethical and political deliberation to offer useful advice and avoid the problems with neutrality, but we must recognize that they will do so despite lacking an adequate claim to representativeness. It would be better for advisers to consider the interests and values of the public and policy makers rather than simply relying on what they personally believe is the ethical thing to do, but we must ultimately accept that expert committees will remain relatively unrepresentative pockets within the institutional landscape of democracies. Once we admit this, we can focus on increasing the democratic legitimacy of these bodies through other means.

Scientific Dissent and Public Scrutiny

The discussion of recent proposals for scientific advice reinforces our main dilemma: proposals that emphasize neutrality compromise their usefulness, while those that aim for more useful advice end up giving scientists a political role that they are ill equipped to fulfill. There is simply a limit to how satisfactorily this dilemma can be solved at the committee level. I therefore propose that we try to mitigate the problem by changing the scope of the issue: instead of asking how scientists could respond to these contradictory demands within a committee, we should focus on how the inherent limitations of advice might be addressed through broader political processes. We should conceive of the scientific advisory committee as initiating and guiding broader democratic debate over science rather than settling the science for policy makers. This would remove the pressure on committees to artificially separate the facts from the values, while reducing the stakes on their necessarily limited attempts at representing diverse societal interests.

The argument for broadening the scope of the issue is primarily on the democratic grounds of more inclusion, better representation, and increased accountability rather than ensuring good outcomes defined independently of democratic procedures. Still, it is likely that a more inclusive public debate, with participation from the rest of the scientific community as well as affected citizens, NGOs, and activists, would be more effective in examining the assumptions of the committee, questioning its mapping of facts and values, and articulating a broader range of values and perspectives than members of the committee alone. While the desirability of democratic debate over scientific advice should not be controversial, what is distinctive about my argument is the suggestion that this aim should guide the design of scientific advice procedures. Scientific advice is usually handled within elite channels, and accepted (or dismissed) without much scrutiny and debate. Its internal dynamics are typically examined independently of broader political processes. For instance, for many years, the Intergovernmental Panel on Climate Change responded to pressures for public accountability mainly in terms of more effective communication strategies, failing to consider how its decision procedures could be revised to engage with the concerns of democratic publics around the world.[37] It remains unusual to think of advisory committee practices as addressed to a large audience that is expected to take on an active and critical role rather than passively absorbing recommendations.

The suggestion to submit scientific advice to public scrutiny directs our attention to the question of how scientific advice could be structured to facilitate public debate and scrutiny under conditions of asymmetrical knowledge. I will focus on two different ways that this can be done. The first is by integrating scientific advice more directly with general currents of democratic discussion and activism in civil society. In the rest of this chapter, I will discuss a practice that can be adopted at the level of the scientific advisory committee to facilitate this. The second is to explore the use of small-scale democratic

---

37. Silke Beck, "Between Tribalism and Trust: The IPCC under the 'Public Microscope,'" *Nature and Culture* 7, no. 2 (2012): 151–73.

experiments to submit expert advice to scrutiny by groups of ordinary citizens. The next chapter will develop a proposal for such an institution.

For democratic scrutiny to be possible, nonexperts must acquire a sense of the committee's assumptions and priorities as well as the role of uncertainty, disagreement, and value trade-offs in their recommendations. Scientists, however, may not be able to identify their own value judgments as value judgments and might see their recommendations as following closely from scientific findings. While committees must explain and justify their decisions, neither explanation nor justification is enough for accountability; it must also be possible for others to be able to judge the validity of the explanations and justifications offered. The complexity of scientific advice renders this difficult. It can be particularly challenging for nonexperts to determine how relying on other values and ends, or making different normative assumptions, would change the substance of the advice.

The standard practice of aiming for consensus within scientific committees exacerbates this problem.[38] The presentation of a single committee position at the end hides the process from view and erases the alternative viewpoints that were considered but ultimately discarded. This makes it difficult for others to appreciate the weaknesses of the committee's advice, and envision objections and alternatives. The fact that there were different views on a committee and the consensus was the result of a decision procedure instead of uncoordinated scientific convergence is a crucial piece of information that should be visible to the public.

Moore proposes that committees should disclose the results of their votes in order to signal that there were different views. This proposal, which would have been anathema to the *Ponts-et-Chaussées* engineers, shows the right logic for holding experts accountable, but does not go far enough. A record of committee votes does not reveal much about the sources of disagreement, and how reasonable or significant they were. I suggest that we push this idea toward a more radical conclusion and enhance the accountability of scientific committees by adopting a practice from the US Supreme Court: the writing of

38. Moore, *Critical Elitism*, 134–36. See also David Guston, "On Consensus and Voting in Science: From Asilomar to the National Toxicology Program," in *The New Political Sociology of Science: Institutions, Networks, and Power*, ed. Scott Frickel and Kelly Moore (Madison: University of Wisconsin Press, 2006), 378–404; Philippe Urfalino, "Reasons and Preferences in Medicine Evaluation Committees," in *Collective Wisdom: Principles and Mechanisms*, ed. Hélène Landemore and Jon Elster (Cambridge: Cambridge University Press, 2012), 173–202.

dissenting opinions. In the past three decades, Supreme Court judges have routinized the practice of offering dissenting minority opinions on rulings.[39] These opinions not only record disagreement with the decision but also enhance the broader social and democratic role of the court's decisions.[40] If court decisions embody the authority and finality of law—and sometimes scientific expertise as well—dissenting opinions open these up to scrutiny in democratic processes that rest on the opposite idea: that the possibility for revision and change always remains open. Since the court's verdict is binding on the litigants, the real impact of a dissent is on these democratic processes outside the courtroom.

The increasing prevalence of Supreme Court dissents in the United States over the past century reflects a change in the understanding of the meaning of the court's decisions, from statements of fixed and immutable principles to flexible and revisable decisions for a particular time and social purpose.[41] This closely resembles the view of scientific advice that I have defended here: as uncertain, fallible, and contested recommendations intended for a particular purpose, rather than statements of certain and timeless scientific facts. It thus makes sense for scientific advisory committees to adopt the court's practice of offering one or more dissenting opinions, thereby explaining and defending alternatives that were rejected in the committee. Such dissenting opinions on scientific advisory committees would have both epistemic and democratic value.

The main epistemic value would lie in recording and keeping alive the views that lost out. The pressure for simplicity and agreement pushes committees toward settling on evidentially well-supported options that might be highly diluted; committees tend to converge on the least common denominator.[42] I mentioned earlier that this tendency creates an informational loss that might increase the chances of error and rule out the pursuit of certain ends. It is possible that other citizens will find alternatives discarded by a committee more significant and useful. The awareness that dissenters might write separate opinions would also improve the majority view by encouraging more attention to

39. On this practice, see Robert Post, "The Supreme Court Opinion as Institutional Practice: Dissent, Legal Scholarship, and Decisionmaking in the Taft Court," *Minnesota Law Review* 85 (2000): 1267–391; William J. Brennan, "In Defense of Dissents." *Hastings Law Journal* 37 (1985): 427–39; Lani Guinier, "Demosprudence through Dissent," *Harvard Law Review* 122 (2008): 6–137.

40. Guinier, "Demosprudence through Dissent."

41. Post, "The Supreme Court Opinion as Institutional Practice."

42. Oppenheimer et al., *Discerning Experts*. Recall that Betz has promoted this as an effective way to protect scientists' neutrality. See Betz, "In Defence of the Value Free Ideal."

the limits and uncertainty of arguments and evidence, and more careful consideration of the assumptions underlying their conclusions. After all, scientists on the committee would have the clearest understanding of the assumptions, uncertainty, and possible error of its conclusions. The possibility that these would be publicly exposed in a dissenting opinion would be a disciplining force ensuring that committee reports are well supported, and refrain from overstating or understating the uncertainty of the evidence. This would improve the committee's advice.

The democratic value of dissent would likewise be significant. The expression of divergent views from the committee would facilitate critical scrutiny in the public sphere. Nonexperts would have a better chance of examining expert views if they had guidance from experts themselves. Having several opinions from a committee would support dissenting views in society and provide stronger scientific grounds for dissent where such grounds can be found. It would also provide assurance that important alternatives have not been suppressed. The depth and breadth of the written dissent would reveal crucial information about how settled the scientific opinion is, which would be conveyed more persuasively through a dissent than through a single report. Activists or social movements could find arguments and support from accessibly written dissenting views. Dissents would allow the pursuit of different policy strategies at different levels of decision-making too, thus facilitating experimentation.

While it may be impossible to fully resolve the tension between neutrality and usefulness at the committee level, the presentation of majority and minority opinions strikes a balance between the two. It contains more useful information than the presentation of purely scientific information, as it involves judgments about the sufficiency, significance, and relevance of the evidence. The fact that there is a majority opinion gives a useful signal to decision makers and the public about the overall direction of opinion on the committee. At the same time, the presentation of dissenting views retains some of the advantages of a neutral presentation of different possibilities: it expands the scope for choice, clarifies the limitations of each view, and presents alternative ways to map the facts onto values. The writing of dissenting opinions would be the first step to broader processes of scrutiny. It would initiate debate and questioning rather than settling it. The aim would be to trigger both formal processes of scrutiny by authorized officials, such as hearings, open floor debates, and civil society consultations, and informal processes of public opinion formation, discussion, and resistance in civil society.

Science and the Media

Scientific advice cannot reach a truly public audience unless it is mediated by the media. At first glance, the media appears to face a dilemma similar to that of the advisory committee. Journalists must provide useful information to the public, selecting what is important and simplifying complex information. They are also expected to present information neutrally and refrain from taking sides. The problem, once again, is that these two pressures are incompatible. The judgments required for usefulness and accessibility—judgments about the quality, significance, relevance, and meaning of complex information— invite the charge of nonneutrality. Meanwhile, efforts to maintain neutrality through a balanced presentation of "both sides" of an issue or practices that conceal the journalist's opinion can end up misleading readers as well as obscuring the truth. In both cases, journalists might fail to fulfill their democratic role—either by usurping citizens' ability to reach independent judgments or failing to guide it.

Despite this basic similarity between the democratic dilemmas of scientific advice and journalism, there are important differences between the two, making the journalist's problem less acute. First, the media's claim to neutrality is even less convincing than claims to scientific neutrality. Any selection, framing, ordering, and interpretation of facts will represent a particular perspective. While accuracy, truthfulness, and fairness are reasonable values for journalists to aim for, neutrality is less coherent even as an ideal. Second, while the government usually has a single official advisory body on any scientific issue, the media landscape is diverse and competitive. Different media organizations represent different viewpoints, and citizens can obtain information from many sources. It is not intrinsically problematic for each media organization to represent one perspective, as long as a diversity of viewpoints is maintained across accessible sources.[43]

Journalists can therefore prioritize usefulness over neutrality and aim to provide information in ways that will help citizens reach informed judgments in light of their own values. In the context of scientific advice, the most useful thing that journalists can do is to initiate and exemplify the critical scrutiny of

---

43. This was one justification for ending the Federal Communications Commission's fairness doctrine. See Lee C. Bollinger, "The Rationale of Public Regulation of the Media," in *Democracy and the Mass Media*, ed. Judith Lichtenberg (Cambridge: Cambridge University Press, 1990), 355–68.

expertise that I have argued for, rather than relaying information uncritically, and without attention to the needs and values of citizens. This would involve emphasizing the uncertainty and limits of new scientific findings, pointing out the value assumptions made by scientists, and drawing attention to value trade-offs that scientists may not have considered. These require attention to techniques for interpreting and communicating uncertainty effectively—numerically, visually, and verbally—as well as the ability to make good second-order judgments about the quality of different sources of evidence. Advisory committee dissenting opinions or public statements from other scientists on the limits of the advisory committee views would help journalists in this task. Journalists can use scientific dissents to clarify the limits of different views and expose their background assumptions.

The greatest value of this role would be in modeling the processes of critical reasoning that all citizens should apply to scientific advice, and normalizing the idea that science can be uncertain, incomplete, and yet still useful. Journalists would be more effective in demonstrating a process of scrutiny than in declaring winners and settling scientific disputes. The trained judgment of journalists—and especially science journalists—in dealing with evidence and scientific sources makes them well suited to the role of clarifying the quality and limits of scientific views for other citizens. It might be helpful for journalists to adopt particular perspectives from which to examine sources and criticize authorities, since all judgments about the meaning, significance, and sufficiency of the evidence require assumptions about the interests and needs of others. This does not amount to biased reporting, which involves presenting information in ways that favor certain interests over others.

My emphasis on scientific uncertainty and disagreement may raise the worry that this amounts to a defense of two-sides journalism, which frames all issues as a conflict between two sides and often creates a false equivalence in the process. The use of this structure in climate change reporting, for instance, resulted in disproportionate attention being devoted to denialists and distorted public perceptions of the state of climate science. Neither the journalistic impulse to present different sides of an issue nor even the convention of selecting only two sides is to blame, though. The problem arises from the application of this conventional structure without sufficient reflection on the credibility and evidentiary support of the sides, and without good justifications for selecting those two sides.

The inclusion of two rival viewpoints in a news story cannot achieve neutrality; its relationship to neutrality is mostly performative. This presentation style, however, does convey the message that the two views selected are those that matter—scientifically and politically. Journalists must therefore take responsibility for their judgments about which views they think matter the most for the public and then demonstrate how people can think about these views critically. This does not preclude covering dissenting opinions, especially on issues where the science is highly uncertain and rapidly changing, as long as journalists examine the content, quality, and credibility of dissenting and majority views alike. Dissenters who do not engage with objections and refuse to change their positions in light of new evidence, however, should not be given further coverage. The trained judgment of journalists is invaluable for these tasks.

The widespread dissemination of misinformation in the media landscape today poses a challenge for these efforts. It is critical for science and democracy alike to find strategies to counteract misinformation campaigns. To this end, scientific organizations themselves could develop online strategies and platforms—for instance, by setting up operations to monitor networks and websites that spread false scientific information, and responding through rebuttal campaigns on digital and social media.[44]

Objections

One objection I anticipate is that a model encouraging dissenting opinions and broader democratic scrutiny would undermine trust in science while giving politicians leeway to do whatever furthers their political agenda. Politicians would get away more easily with choosing a dissent over the majority opinion than with ignoring a consensus report. To prevent politicians from abusing scientific advice, scientists should resolve their technical disagreements internally, and present a consensus view to policy makers and the public.

It is easy to see why this view may appear plausible given recent examples of politicians ignoring or denying scientific advice, and it is difficult to refute it completely without empirical evidence on how policy makers and scientists behave under different advisory arrangements. This evidence is difficult to obtain because it is hard to know when politicians follow or reject advice in good faith rather than ulterior motives, and because it is not common for scientific advisory committees to offer public dissents. It is generally difficult to make all-things-told assessments of the expected consequences of institutional

44. Shanto Iyengar and Douglas S. Massey, "Scientific Communication in a Post-Truth Society," *Proceedings of the National Academy of Sciences* 116, no. 16 (2019): 7656–61.

recommendations without empirical evidence, which we cannot have without testing the proposal. In the absence of this evidence, I will offer some reasons why this objection is not convincing.

The plausibility of the objection rests on a specific set of assumptions about science, experts, and politicians, several of which I have been arguing against in this chapter and the last. The claim that it would be best for scientists to present consensus views would be most persuasive if the consensus of the committee were likely to be true, and it either relied on no assumptions about moral and political matters, or else the "right" assumptions.[45] I have been arguing that these cannot be assumed because of the intrinsic uncertainty and incompleteness of science and the role of scientists' own values in shaping scientific advice. Might we have reason to prefer consensual arrangements and discourage dissent even once we grant the uncertainty, incompleteness, and value ladenness of science? Perhaps, if we assume that politicians will ignore the science and the interests of the public, but even this assumption alone will not be enough. We also need to assume that scientists could be trusted more than politicians to know and be motivated to advance the right political aims, and once again that the consensus view of the committee is unlikely to be mistaken. These three assumptions together, while not impossible to meet, are unrealistically asymmetrical in their level of idealization. They assume the worst of politicians and the best of science and scientists. Even if we grant pessimism about politicians, we should resist idealizing science and scientists at the same time.

I have already argued that a small group of scientists could not adequately represent the range of interests and views in society even though some representation must happen on the committee. There are no channels through which scientists could discharge this task, and none to hold them accountable if they made mistakes in their judgment. They will inevitably fall back on their own personal judgment, which is unlikely to be representative. Electoral politics, by contrast, is designed to achieve representation of and accountability

---

45. Philosopher Stephen John's provocative argument against honesty in science communication, for instance, assumes that the nonexpert ought to defer to claims that meet standards of scientific acceptance, even if the scientific consensus is artificial (i.e., reached through a vote). On this view, the aim of scientific communication is to secure the deference of the nonexpert. I disagree with this basic assumption, especially given the role of value judgments, uncertainty, and disagreement in science. Stephen John, "Epistemic Trust and the Ethics of Science Communication: Against Transparency, Openness, Sincerity and Honesty," *Social Epistemology* 32, no. 2 (2018): 75–87.

to the public—however inadequately it works in practice. While it is hard to deny that politicians can get away with ignoring scientific advice, the solution to the problems of electoral politics should not be to thrust scientific advice into a political role it is not fit for. Encouraging and facilitating democratic scrutiny is a way to hold scientists accountable, and facilitate the representation of a broader spectrum of interests and values.

To shake off the intuition that consensual advisory processes are more desirable as a rule, I will examine a case that illustrates how things can go wrong when scientific advice is unanimous, authoritative, and mistaken. In February 1976, there was a small outbreak of swine flu among army recruits at Fort Dix, New Jersey. One soldier died. The United States had not experienced a swine flu outbreak since 1918–19, and the possibility of another pandemic raised alarm. The Centers for Disease Control (CDC) called a special meeting of its Advisory Committee on Immunization Practices to consider the evidence and make recommendations. There was no evidence of outbreaks elsewhere in the country, and scientists could not rule out the possibility that small-scale outbreaks had been occurring undetected without leading to a pandemic.[46] It was also unclear how virulent this strain of swine flu really was. Predictions about the likelihood and severity of a pandemic were highly conjectural.[47] Several scientists on the committee did not believe that there was a real threat. The committee initially recommended the production of a vaccine, but stopped short of suggesting its administration.

The CDC director, however, was persuaded of the need to go ahead with a mass immunization program. He conducted a telephone poll of committee members, asking each of them whether they would oppose the recommendation that all citizens be immunized within three months. The director gained the assent or acquiescence of all the members and reported a unanimous decision.[48] His memo moved swiftly through the bureaucracy and was accepted by President Gerald Ford, who announced the program to vaccinate "every man, woman, and child in the United States." By the end of the year, forty million people had been vaccinated at a cost of $135 million. Yet the pandemic never materialized, and the vaccine turned out to be associated with an increased risk

46. Philip M. Boffey, "Anatomy of a Decision: How the Nation Declared War on Swine Flu," *Science* 192, no. 4240 (May 1976): 636–41.

47. Richard E. Neustadt and Harvey V. Fineberg, *The Swine Flu Affair: Decision Making on a Slippery Disease* (Washington, DC: National Academies Press, 1978).

48. Boffey, "Anatomy of a Decision"; Neustadt and Fineberg, *The Swine Flu Affair*.

of Guillain-Barré syndrome—a rare nervous system disease. Perhaps the greater damage was that the incident reduced citizens' trust in the value and safety of public immunization programs and government public health initiatives for years. The CDC director was fired, and the incident was widely regarded as a fiasco.

It is not surprising that the science was uncertain and the best scientific advisers, apparently in good faith, made a mistaken assessment on its basis. What is distinctive about this episode—and science-politics relationships in this period more generally—is the president's readiness to defer to the committee without further questioning or scrutiny. Congress was not consulted before the announcement, and the whole process was conducted in secret within the bureaucracy. A postmortem report of the incident, commissioned by the secretary of the US Department of Health and Human Services, concluded that the scientists were overconfident, and dissent within the committee was suppressed by the desire to present a consensus view.[49] Several scientists who served on the CDC advisory committee later expressed their skepticism, claiming that they thought the chance of a pandemic was small and they did not believe going ahead with an immunization program was justified, but this dissent had not been conveyed to policy makers.[50] The contrived telephone polling is a reminder that decision makers may know little about how unanimity is reached on a committee.

My argument here is not driven by the strategic aim of maximizing trust in science. I think the amount of trust warranted in scientific advice cannot be determined without scrutiny and reflection on a case-by-case basis, especially given uncertainty, fallibility, and the role of values. Still, it is worth considering whether the critical mechanisms that I propose on democratic grounds would have such negative effects on the credibility of scientists that this worry ought to trump any possible democratic gains from scrutiny. The swine flu debacle provides one strong counterpoint to this. While it supports the view that a unanimous scientific opinion is more difficult for politicians to ignore, it also illustrates the possibility that mistakes from respected advisers can inflict significant and lasting damage to public trust in science. Trust takes a long time to build, but can be destroyed instantly.[51] Moreover, a reduction of public trust in

49. Neustadt and Fineberg, *The Swine Flu Affair*.

50. Boffey, "Anatomy of a Decision."

51. Paul Slovic, "Perceived Risk, Trust, and Democracy," *Risk Analysis* 13, no. 6 (1993): 675–82.

response to a scientific fiasco is not only inevitable but fully warranted too. Efforts to manage public trust through strategic disclosures will backfire when scientific advice is mistaken. Since people with different values will reasonably regard different conclusions to be warranted on the basis of the same evidence, there will always be a possibility of backfire for some subset of the population.

The objection about trust can be challenged on empirical grounds as well. This worry assumes that trust in science is bolstered by authoritative and certain scientific assertions, and diminished by admissions of uncertainty and disagreement. It turns out, however, that this common view of the inverse relationship between uncertainty and trust is not supported by empirical evidence. Recent work on communicating scientific uncertainty and risk has shown that people do not reduce their trust in scientific findings if uncertainty is reported, especially if the uncertainty is expressed numerically rather than verbally.[52] Findings are robust across different sources and types of uncertainty. It is important to remember that we are not just considering the accuracy of science but also the value judgments made on a committee on its basis. There will always be some people who would have made different value judgments on the basis of the same level of evidence and uncertainty, and they would have more reason to trust the committee if they knew that their concerns were considered.

Another question (if not quite an objection) that I anticipate is whether the suggestion for advisory committees to write dissenting opinions is merely a call for greater transparency. On the surface, this proposal is of course related to transparency; it is a demand for committees to share more information about their beliefs and disagreements with the public. This would be valuable for all the reasons that transparency is valuable: it would prevent the misrepresentation of advice to the public, make shifting blame to experts more difficult, and make it easier to expose government officials' false claims to be following the science. But on a more nuanced level, my proposal requires both less and more than transparency, understood as the disclosure of information. It does not necessitate complete transparency because it does not involve the disclosure of internal deliberations or the immediate release of meeting minutes. Political theorists have shown that secrecy can be valuable in deliberations within small groups, allowing participants to air more controversial claims and

52. Anne Marthe van der Bles, Sander van der Linden, Alexandra L. J. Freeman, and David J. Spiegelhalter, "The Effects of Communicating Uncertainty on Public Trust in Facts and Numbers," *Proceedings of the National Academy of Sciences* 117, no. 14 (2020): 7672–83.

offer candid opinions about the weaknesses of their own positions.[53] These can improve the quality of the resulting advice. The proposal for dissenting opinions requires the disclosure only of unresolved disagreements. Moreover, since dissent would be explicitly directed at a public audience, it would include only the information that dissenters think the public and policy makers ought to know.

At the same time, this proposal requires something that cannot be reduced to mere transparency: a culture of criticism and dissent within a group of advisers, and willingness to speak to the public. This, in turn, requires the cultivation of professional norms that make it acceptable to express disagreement and do so publicly as well as the adoption of formal rules that permit majoritarian decision-making and written dissent in committees. None of these would be meaningful if a diversity of viewpoints could not be found on the committee in the first place. These circumstances cannot be taken for granted. Advisory committees often have an incentive to hide their disagreements, whether because they believe disagreement indicates scientific error or simply to bolster their political authority. Both the swine flu advisory group and *Ponts-et-Chaussées* engineers were intent on making their decisions unanimously, however that might be achieved. Politicians might prefer unanimous decisions too in order to shift the blame for consequences of policies adopted on the basis of scientific advice. The swine flu case illustrates that transparency alone would not provide valuable information about the limits of the committee advice if dissent were suppressed or discouraged within the committee. Scientists who admitted their skepticism during the official inquiry afterward had acquiesced to the director's recommendation because of the pressure to make unanimous decisions. In the absence of the norms and rules of criticism, transparency alone would not be enough.

Returning to the *Ponts-et-Chaussées*

Let me return to the paradox that I started out with and assess where we have ended up with respect to it. I argued that there is an inevitable trade-off between the neutrality and usefulness of scientific advice, and that advisory

---

53. Jonathan Bruno, "*Democracy beyond Disclosure: Secrecy, Transparency, and the Logic of Self-Government*" (PhD diss., Harvard University, 2017); Simone Chambers, "Behind Closed Doors: Publicity, Secrecy, and the Quality of Deliberation," *Journal of Political Philosophy* 12, no. 4 (2004): 389–410.

committees tend toward one or the other. I then pointed out serious limitations of the neutrality model in dealing with the interplay of evidentiary and practical considerations in committee advice. Approximating neutrality either masks implicit value judgments or risks irrelevance by falling back on scientific technicalities. I argued that it is therefore preferable to aim for useful advice and accept that certain kinds of value judgments must be made on expert committees.

Still, moving away from efforts to maintain a strict demarcation between facts and values opens up a Pandora's box of concerns around democratic representation. The need for value judgments on scientific committees means that scientists must deliberate about matters that fall outside their areas of competence, and on which they will be no better informed or qualified than nonexperts. Since the spectrum of political viewpoints cannot be adequately represented on an expert committee, scientific advice could be biased, narrow, and lacking important perspectives and values. This concern is more serious for committees at the forefront of new research that set a new political agenda rather than responding to clearly defined social ends and values.

This diagnosis of the political challenges of scientific advice is the starting point for efforts to prescribe institutional measures in response. To mitigate the difficulties of representation on scientific committees, I argued that mechanisms of ex post democratic accountability must be strengthened in ways that allow science to be closely scrutinized and contested by nonexperts before being accepted as the basis of policy decisions. I also suggested that scientific advisory committees could facilitate such scrutiny by offering both majority views and dissenting opinions in their reports, and the media could have a key role to play in directing citizens to important views and demonstrating how to think critically about them. I will continue to explore the possibilities for the democratic scrutiny of expertise in the next chapter, where I develop an institutional proposal for a science court.

As you might recall, the engineers of the *Ponts-et-Chaussées* had resisted the vote because they believed purely technical disagreements should be resolved through continued technical discussion. As one engineer put it, they were convinced that starting from clear principles and "following the simple laws of logic" would ensure that "the inferences drawn . . . could not be contested by anyone."[54] The argument in this chapter suggests that instead of a purely technical controversy, the engineers were faced with a practical decision under

54. Graber, "Obvious Decisions," 945.

uncertainty, requiring them to weigh various practical considerations along with the evidence. The decision to settle on certain views about the safety, navigability, feasibility, and cost of the canal required considering the risk of being wrong in a variety of ways (especially in the case of the more ambitious Laurent project) as well as making judgments about the relative goodness of potential consequences and sufficiency of the evidence for different purposes. When the case is reinterpreted this way, the vote no longer appears to be an irrational procedure for making the decision; in fact, prolonging the technical deliberation appears to have been the misguided move.

But this interpretation of the case raises doubts about whether the engineering corps was best placed to make the ultimate choice between the canal projects. After all, the assembly was faced with practical considerations and trade-offs that did not follow in any straightforward way from the evidence at its disposal. Interestingly, the engineers sensed that they were perhaps not the right ones to make the judgment, so they decided early on to present a detailed summary of their assessment and turn over the decision to Bonaparte. Better still, they offered the government not only the majority view but also a lengthy minority opinion. It was Bonaparte who refused to make the decision and insisted that the engineers should decide instead.

The twist in the story is that Bonaparte later became irritated with the assembly's inability to decide and its stubborn aversion to the vote, so he unilaterally reversed the assembly's final decision. It might be concerning that Bonaparte's petulant reversal of the engineers' decision was, at least on the surface, indistinguishable from a reflective judgment that the canal project supported by the minority was in fact the better one. Politicians who shirk the responsibility to scrutinize expert advice are unfortunately all too common, then as now. This is not an argument against the need for the democratic scrutiny of science but rather a warning against unconstrained and unaccountable politicians. The next chapter takes this problem seriously and develops a proposal for ordinary citizens to become directly involved on policy issues requiring complex scientific knowledge.

4

A Proposal for a Science Court

IN A SERIES OF ARTICLES written in the 1960s and 1970s, physicist Arthur Kantrowitz developed a proposal for a new institution for dealing with controversial scientific issues in policy making. The proposal was designed to address the problem of expert disagreement on scientific issues that required political decisions, ranging from nuclear power and the ozone layer to food additives and fluoridation. Kantrowitz lamented the state of public debate over scientific controversies. He complained that competing experts made contradictory technical claims in the public sphere that did not get challenged or refuted directly. This left the public in confusion about the state of scientific knowledge, weakened the scientific basis of public policy, and heightened mistrust of experts.[1]

His solution was to create an adversarial institution in which rival experts would defend their case and then cross-examine each other in front of a panel of impartial scientist judges. The judges would then reach a verdict on the disputed scientific points and highlight points of agreement between the two sides. The proceedings would be open to the public, and the decision would serve an advisory role for Congress and the president. Kantrowitz initially called his proposal an "institution for scientific judgment," but the media coined the pithier term "science court," which stuck.[2]

---

1. Arthur Kantrowitz, "Proposal for an Institution of Scientific Judgment," *Science* 156, no. 3776 (1967): 763–64; Arthur Kantrowitz, "The Test: Meeting the Challenge of New Technology," *Bulletin of the Atomic Scientists* 25, no. 9 (1969): 20–22, 48; Arthur Kantrowitz, "Controlling Technology Democratically," *American Scientist* 63 (1975): 505–9; Arthur Kantrowitz, "The Science Court Experiment," *Jurimetrics Journal* 17 (1977): 332–41.

2. Andrew Jurs, "Science Court: Past Proposals, Current Considerations, and a Suggested Structure," *Virginia Journal of Law and Technology* 15, no. 1 (2010): 1–43.

The proposal had three key features. First, it would separate the facts and values involved in scientific controversies, and then evaluate only the facts. Second, it would separate the advocate and judge, and use adversary proceedings. The third rule was that the judges had to be scientists, although they were not supposed to be specialists on the question being judged. These three features together were meant to ensure the objectivity and accuracy of science advice.

By 1975, the science court proposal had acquired great popularity and was even supported by the White House.[3] President Ford created a task force of academics and government officials within his Advisory Group on Anticipated Advances in Science and Technology to explore the feasibility of the proposal. Kantrowitz was appointed chair. The task force decided to run a series of preliminary experiments to test the science court, and evaluate its benefits and drawbacks.[4] Two hundred and fifty scientists and legal scholars participated in a public debate organized to discuss the proposal, and despite some criticism, many of the points in the proposal met with approval.[5] Twenty-eight prominent scientific organizations, including the American Association for the Advancement of Science, offered their support.[6]

An opportunity for testing the science court came from Minnesota in 1976.[7] There was a controversy over the construction of a high-voltage power line that would cut across 172 miles of farmland. The farmers were deeply upset, not only because their lands would be appropriated, but because they believed that the selected path would be particularly harmful for irrigation patterns and other farming practices. They also had concerns about health and safety issues along with the potential environmental damage that would be caused by the power line. The utility companies, however, denied that the farmers' claims had any scientific basis. The governor of Minnesota stepped in and proposed to resolve the scientific aspects of this dispute in a science court.

3. Jurs, "Science Court."

4. Philip M. Boffey, "Experiment Planned to Test Feasibility of a 'Science Court,'" *Science* 193, no. 4248 (July 1976): 129; John Noble Wilford, "Science Considers Its Own 'Court,'" *New York Times*, February 29, 1976, 140.

5. Boffey, "Experiment Planned to Test Feasibility of a 'Science Court'"; Wil Lepkowski, "USA: Science Court on Guard," *Nature* 263 (1976): 454–55.

6. John Noble Wilford, "Leaders Endorse Science Court Test," *New York Times*, January 2, 1977, 28.

7. Barry Casper and Paul Wellstone, "The Science Court on Trial in Minnesota," *Hastings Center Report* 8, no. 4 (1978): 5–7.

He took on the responsibility of organizing it, and tried to persuade the farmers and utility companies to participate.

But the farmers refused to participate in a science court under the rules proposed by the governor—basically the rules of the Kantrowitz proposal. They saw the court's separation of the factual and political aspects of the problem as a cover for delegating an essentially political decision-making power to technical experts. Instead, they proposed a modified court where the scientific and political parts would be argued together. Of course, if the facts and values were to be addressed together, it no longer made sense for the judges to be scientists, so the farmers asked the governor himself to act as judge. Their two other demands were for funding to develop their case and bring in their own experts, and the hearings to be directed at the public.

In the end, the governor rejected the farmers' demands, and the science court never took place. All this happened during a presidential election year. Although both candidates, Ford and Jimmy Carter, had publicly endorsed the science court in their campaigns, thus proving and increasing the popularity of the idea, Carter completely abandoned the project when he was elected president. The idea silently disappeared.

This chapter develops a proposal for reviving the science court, but in a form closer to the one proposed by the farmers. My proposal is intended to illustrate one way the democratic scrutiny of expertise could be institutionalized alongside formal scrutiny by elected officials, critical reporting by the media, and informal processes of debate in the public sphere. The most distinctive aspect of my proposal for a science court is that it involves direct participation by randomly selected groups of citizens. While I argue that citizen involvement is desirable on democratic grounds, there are well-known obstacles to meaningful participation on issues involving expertise given conditions of unequal knowledge, authority, and credibility. Ordinary citizens cannot set the agenda and terms of the debate; they face challenges in evaluating complex technical claims; and asymmetries in knowledge and authority make deliberation between experts and citizens unproductive.

To respond to these three challenges, I suggest an adversarial institution that can be initiated by citizens, and where competing experts make the case for different positions on a policy question with a significant scientific component. A citizen jury (instead of Kantrowitz's scientist judges) interrogates the experts, then deliberates and delivers a decision, evaluating the facts and their practical implications together. The adversary structure of my proposal is intended to expose the background assumptions, potential biases, and omissions in rival expert claims as well as clarify the levels of uncertainty. The separation of scientist advocates from citizen jurors avoids the difficulties of mutual deliberation under conditions of unequal authority, while placing citizens in the seat of judgment.

The Limits of Parliamentary Scrutiny and Existing Minipublics

The last two chapters have argued that scientific advice must be submitted to democratic scrutiny. There are many ways such scrutiny can happen, but the default solution in a representative democracy is to assign the responsibility to political representatives, and rely on parliamentary resources and procedures to facilitate the proper scrutiny of expertise. This replaces the division of labor between experts and nonexperts with an elite-nonelite one. Representatives examine expert claims, and make sure that the values and purposes of experts do not drive their advice, while ordinary citizens concern themselves only with deliberation over ends. Thomas Christiano provides the clearest defense of this model. He argues that citizens must deliberate on the ultimate aims of society, and representatives must be responsive to these deliberations, but "the citizen is not expected to have an understanding of the specialized knowledge the other persons have."[8] Political parties, legislators, and special interest groups can be entrusted with the proper handling of expert advice to select the appropriate means for realizing citizens' aims.

While Christiano is right that representatives bear the primary responsibility for consulting experts and evaluating scientific advice for policy purposes, leaving the task entirely to them cannot fully address the problems with expertise discussed so far and creates some additional difficulties. One problem is that the ends and means are not as clearly separable as Christiano assumes.[9] To judge whether representatives are responding to their values and preferences, citizens must have a sense of what courses of action are possible, to what extent the available evidence supports them, and whether the evidence

8. Thomas Christiano, "Rational Deliberation among Experts and Citizens," in *Deliberative Systems: Deliberative Democracy at the Large Scale,* ed. John Parkinson and Jane Mansbridge (Cambridge: Cambridge University Press, 2012), 42; Thomas Christiano, *The Rule of the Many: Fundamental Issues in Democratic Theory* (Boulder, CO: Westview Press, 1996).

9. Henry S. Richardson, *Democratic Autonomy: Public Reasoning about the Ends of Policy* (New York: Oxford University Press, 2002).

is sufficient to justify the action ultimately taken. To evaluate whether the potential benefits from a policy are worth the risk of harm created, citizens must know the likelihood of benefits and harms as well as the uncertainty and completeness of these assessments. If bad policies are pursued on the grounds that the evidence supports it, citizens must be able to determine if the science is biased or else blame their representatives for acting in bad faith. Citizens cannot set the aims of society, as Christiano correctly argues that they should, without some ability to judge the scientific evidence claimed in support of policy decisions. Opportunities for examining the means as well as the ends is crucial for citizens to hold decision makers accountable.

Citizens can choose to trust their representatives, of course, and can also trust experts. Each citizen could not and need not scrutinize the scientific basis of each policy decision. Nevertheless, a division of labor built on the assumption that citizens as a whole should not engage with the expertise used as the basis of policies would be problematic. A system of representation that runs on the justified trust of officials or experts most of the time must ultimately be predicated on the existence and proper functioning of underlying mechanisms of accountability—that is, distrust.[10] In policy areas that rely heavily on expertise, the usual channels of accountability through publicity will not be effective because citizens will lack the knowledge to scrutinize and judge policy decisions, however transparent they may be. In other words, the problem is not only that citizens will lack the effective ability to set the aims of policy and influence political decisions directly, though these are intrinsically problematic from a democratic point of view. It is that this inability will encourage irresponsible decision-making by elites. If citizens cannot scrutinize the factual component of policy decisions, and this inability is known among representatives and experts, the advisory relationship is likely to be corrupted at the expense of the public interest.

Let me describe how things could go wrong. First, experts might go beyond their advisory role and act as policy advocates, and legislators might allow this overreach if it suits their own political interest to let "the science" compel a certain line of action. They might leave the hard choices and value trade-offs to experts, while making symbolic statements at the level of generalities.

---

10. Jonathan Bruno, "*Democracy beyond Disclosure: Secrecy, Transparency, and the Logic of Self-Government*" (PhD diss., Harvard University, 2017); Mark E. Warren, "Democratic Theory and Trust," In *Democracy and Trust*, ed. Mark E. Warren (Cambridge: Cambridge University Press, 1999), 310–45.

Democratic politics might thus turn into rule by experts. Another possibility is that well-meaning experts may try to be neutral among alternative courses of action, and the process of translating the knowledge into a concrete policy might be co-opted by advocacy groups and special interests trying to push their own narrow agenda. Since representatives know that the public cannot evaluate the knowledge that forms the basis of a policy, they will be more likely to respond to powerful sectional interests rather than pursuing policies in the public interest. Representatives might also simply ignore the advice of experts, misrepresent it to the public, rule out certain kinds of advice by their framing of questions, or try to skew the advisory process through expert appointments that favor their personal interests.

These scenarios are not exhaustive, but they illustrate some potential pathologies that could arise if only elites evaluate the factual basis of policy decisions and ordinary citizens are expected to be mere value vessels. The knowledge that people are neither expected nor capable of engaging with the means creates perverse incentives for representatives, and is more likely to result in policies driven by the values and interests of experts and narrow sectional interests. Leaving the means entirely to elites will also make it more likely that representatives become less responsive to public preferences about ends and less troubled about pursuing good policies because they can always claim that they made the best choice in light of the facts. To prevent this, the knowledge base of political decisions must be open to scrutiny by citizens as well, and not just by their representatives.

In the past few decades, scholars have suggested that small-scale participatory experiments that bring together experts and nonexperts to discuss scientific or technical issues could offer a solution to the problem of expertise in democracy.[11] The difficulties of communicating complex information in a

---

11. Robert A. Dahl, *Democracy and Its Critics* (New Haven, CT: Yale University Press, 1989); Ned Crosby, "Citizens' Juries: One Solution for Difficult Environmental Questions," In *Fairness and Competence in Citizen Participation: Evaluating Models for Environmental Discourse*, ed. Ortwin Renn, Thomas Webler, and Peter Wiedemann (Dordrecht: Kluwer Academic Press, 1995), 157–74; James Fishkin, *Democracy and Deliberation* (New Haven, CT: Yale University Press, 1991); James Fishkin, *The Voice of the People: Public Opinion and Democracy* (New Haven, CT: Yale University Press, 1997); Archon Fung, *Empowered Participation: Reinventing Urban Democracy* (Princeton, NJ: Princeton University Press, 2009); John Gastil and Peter Levine, eds., *The Deliberative Democracy Handbook: Strategies for Effective Civic Engagement in the Twenty-First Century* (San Francisco: Jossey-Bass, 2005); Simon Joss and John Durant, eds., *Public Participation in Science: The Role of Consensus Conferences in Europe* (London: Science Museum, 1995).

highly polluted and polarized public sphere have made small, carefully organized institutional venues an attractive option for opening up typically opaque areas of policy to public scrutiny. These so-called minipublics are particularly well suited to deliberation over complex technical issues because they give ordinary citizens direct access to experts, the time and incentive to reflect on the information provided, and the opportunity to engage in face-to-face deliberation, thereby improving understanding and facilitating the critical assessment of the viewpoints presented. They thus combine the democratic qualities of participation and representation with the epistemic advantages of small group deliberation. At their best, institutions such as deliberative polls, citizen juries, consensus conferences, and planning cells hold the potential to strengthen representative democracy by enhancing accountability, increasing participation, improving the quality of decisions, and strengthening democratic influence on representatives between elections.

To be clear, these small-scale democratic experiments cannot be the only conduits for the democratic scrutiny of expertise, especially since they can only involve a small number of people at a time.[12] Party competition organizes and simplifies complex information for the mass public, and shapes the partisan interpretations of scientific findings. The media scrutinizes experts and politicians alike, and directs opinion formation in the public sphere. Advocacy groups and social movements make complex technical issues more salient and visible for the public through contestatory action and resistance.[13] Nonetheless, there are distinct advantages to institutions for face-to-face interaction between experts and laypeople, and that makes it worth considering their role in some detail. Participatory experiments give citizens access to expert information without the partisan filter of parties, and allow for a thorough evaluation of a single issue over several days, weeks, or even months.[14] They

12. Jane Mansbridge, James Bohman, Simone Chambers, Thomas Christiano, Archon Fung, John Parkinson, Dennis F. Thompson, and Mark E. Warren, "A Systemic Approach to Deliberative Democracy," in *Deliberative Systems: Deliberative Democracy at the Large Scale*, ed. John Parkinson and Jane Mansbridge (Cambridge: Cambridge University Press, 2012), 1–26.

13. For a discussion of the role of social movements on issues of expertise, see Alfred Moore, *Critical Elitism: Deliberation, Democracy, and the Problem of Expertise* (Cambridge: Cambridge University Press, 2017), chapter 5.

14. For the epistemic and democratic benefits of single-issue publics more broadly, see Kevin J. Elliott, "Democracy's Pin Factory: Issue Specialization, the Division of Cognitive Labor, and Epistemic Performance," *American Journal of Political Science* 64, no. 2 (2020): 385–97.

also institutionalize an active role for ordinary citizens and ensure a direct channel of influence over policy makers for a cross section of the population that is ordinarily far from influence. I will discuss the relationship between these institutions and broader currents of public opinion later in this chapter. For now, I want to begin with the question of institutional design.

An important problem with existing designs for minipublics is that the selection and presentation of expert information puts inherent limits on participants' ability to contest the knowledge presented. This is because most current deliberative experiments are organized by professionals, who set the agenda, select the material for discussion, invite the experts, and prepare background information.[15] Political scientists have long recognized that those who control the agenda have the power to shape the outcome, and some scholars of deliberative democracy have pointed out that a crucial test for the legitimacy of deliberative settings is the power of citizen groups to initiate the discussion of problems.[16] Existing minipublics have not taken these lessons seriously enough.

Most existing proposals consist of three phases: learning, deliberation, and decision-making. Exposure to experts and expertise takes place at the learning

15. André Bächtiger, Maija Setälä, and Kimmo Grönlund, "Towards a New Era of Deliberative Mini-Publics," in *Deliberative Mini-Publics: Innovating Citizens in the Democratic Process*, ed. Kimmo Grönlund, André Bächtiger, and Maija Setälä (Colchester, UK: ECPR Press, 2014), 225–47; Alexander Bogner, "The Paradox of Participation Experiments," *Science, Technology, and Human Values* 37, no. 5 (2012): 506–27; Kathrin Braun and Susanne Schultz, "'. . . a Certain Amount of Engineering Involved': Constructing the Public in Participatory Governance Arrangements," *Public Understanding of Science* 19, no. 4 (2010): 403–19; Mark B. Brown, "Expertise and Deliberative Democracy," in *Deliberative Democracy: Issues and Cases*, ed. Stephen Elstub and Peter McLaverty (Edinburgh: Edinburgh University Press, 2014), 50–69; Amy Lang, "Agenda-Setting in Deliberative Forums," in *Designing Deliberative Democracy: The British Columbia Citizens' Assembly*, ed. Mark E. Warren and Hilary Pearse (Cambridge: Cambridge University Press, 2008), 85–105; Maria Powell and Mathilde Colin, "Meaningful Citizen Engagement in Science and Technology," *Science Communication* 30, no. 1 (2008): 126–36; Mark E. Warren, "Governance Driven Democratization," *Critical Policy Studies* 3, no. 1 (2009): 3–13.

16. On agenda setting and power, see Peter Bachrach and Morton Baratz, "Two Faces of Power," *American Political Science Review* 56, no. 4 (1962) 947–52; Steven Lukes, *Power: A Radical View* (London: Macmillan, 1974). On agenda setting and deliberative democracy, see James Bohman, "Democracy as Inquiry, Inquiry as Democratic: Pragmatism, Social Science, and the Cognitive Division of Labor," *American Journal of Political Science* 43, no. 2 (1999): 590–607; Iris Marion Young, "Activist Challenges to Deliberative Democracy," *Political Theory*, 29, no. 5 (2001): 670–90.

phase, which usually involves the distribution of background materials to participants, followed by presentations from a range of experts. The problem is that participants are usually passive at this stage, even though this stage has a crucial role in shaping and constraining the subsequent discussion. Both organizers and scholars of minipublics emphasize the importance of providing balanced information at the learning phase, and admit the necessity of offering different perspectives. But they say little about the challenges of producing balance on controversial issues that involve complex scientific knowledge. I argued in the last chapter that summarizing and simplifying are among the most important as well as most value-laden tasks in the use of expertise for policy purposes. All the obstacles to neutrality and the trade-offs between neutrality and usefulness discussed in the context of scientific advisory committees also apply to the process of selecting information for participants in a minipublic. Organizers must make assumptions about the values and priorities of citizen-deliberators or rely on their own priorities to supply useful information. Their choices then shape what the deliberators will consider significant and relevant knowledge. Handing this task entirely to organizers, however independent and competent, takes away a critical power from citizens.

Let me support these claims with a quick survey of the existing proposals. For simplicity, I will classify existing designs for minipublics into two categories based on the part that experts play in each setting. In the first category are minipublics in which experts lecture nonexperts on the technical parts of a problem, followed by citizen groups deliberating among themselves. In the second category are experiments where experts and nonexperts deliberate together to solve a policy problem. The consensus conference is a standard example of the first format.[17] It involves citizens listening to expert testimony, and then deliberating about the social and ethical issues raised by a proposed technological innovation. Its most famous case, the Danish Technology Board, has already come under criticism on the grounds that the steering committee manipulates deliberation through the selection of informational materials.[18] Political scientist James Fishkin's deliberative poll is another well-known example of a deliberative setting that brings together citizens and experts around an educative goal. Expert advice and "balanced briefing materials" are provided

17. Joss and Durant, *Public Participation in Science*.

18. Aviezer Tucker, "Pre-emptive Democracy: Oligarchic Tendencies in Deliberative Democracy," *Political Studies* 56, no. 1 (2008): 127–47.

in order to increase the quality of citizen deliberation. The effectiveness of the experiment is assessed by before and after surveys in which the degree of citizen uptake of information provided by experts counts as a measure of success, making it clear that the organizers hope that the citizens will absorb the expert views.[19]

In one of the most detailed recent treatments of the relationship between science and democracy, Kitcher has put forward a similar proposal. His ideal of "well-ordered science" involves scientists tutoring small groups of citizens about the technical aspects of an issue so that the citizens can go on to deliberate in a more informed fashion. The aim is to transform citizens' preferences by exposure to expert opinion and prevent "vulgar democracy," which Kitcher defines as the shaping of government policy on scientific issues by the "untutored" preferences of citizens.[20] Even the British Columbia Citizens' Assembly, which is unique among deliberative experiments in the degree of decision-making power delegated to it, followed the same educative format for using expertise. The organizers instituted a "learning phase," which consisted of members of the assembly listening to lectures from experts selected by the organizers in light of the mandate given by the government.[21] Interestingly, a study of the assembly showed that the participants were wary about manipulation by the organizers and pushed back against the choices presented to them at several junctures, but the organizers always had the last word.[22]

A widely used alternative arrangement for minipublics involves deliberation between experts and nonexperts.[23] These hybrid deliberative experiments aim to increase participation, accountability, information transfer, and democratic legitimacy, much like educative minipublics, but they are typically more ambitious about empowering people on policy decisions, and often claim that deliberation between experts and nonexperts will increase the

19. Fishkin, The Voice of the People.

20. Philip Kitcher, Science, Truth, and Democracy (Oxford: Oxford University Press, 2001), 117–36.

21. Mark E. Warren and Hilary Pearse, eds., Designing Deliberative Democracy: The British Columbia Citizens' Assembly (Cambridge: Cambridge University Press, 2008).

22. Lang, "Agenda-Setting in Deliberative Forums."

23. Mark B. Brown, Science in Democracy: Expertise, Institutions, and Representation (Cambridge, MA: MIT Press, 2009); Michel Callon, Pierre Lascoumes, and Yannick Barthe, Acting in an Uncertain World: An Essay on Technical Democracy (Cambridge, MA: MIT Press, 2009); Fung, Empowered Participation; Charles Sabel, Archon Fung, and Bradley Karkkainen, Beyond Backyard Environmentalism (Boston: Beacon Press, 2000).

quality of the resulting decisions. For instance, Archon Fung argues that citizens and experts deliberating together could generate better solutions to seemingly intractable technical problems that neither could address satisfactorily on their own.[24]

Minipublics that involve expert-nonexpert deliberation, however, run into problems posed by inequalities in knowledge and epistemic authority. Theories of deliberative democracy have been criticized for setting highly idealized criteria of equality and reciprocity as a precondition for deliberation, and then failing to pay sufficient attention to whether and how these conditions can be met in practice.[25] Critics have pointed out that background inequalities make it extremely difficult for deliberation to be guided by "the unforced force of the better argument" and the results of deliberation under conditions of inequality are likely to be shaped by existing differences in power among the participants.[26] In response to these charges, scholars of deliberative democracy have proposed institutional mechanisms for structuring deliberation in ways that could offset the known effects of background inequalities among the participants.[27] They have also maintained that properly conducted deliberation could be instrumental in mitigating existing inequalities; discussion and argumentation could neutralize the effects of power by exposing it as resting on illegitimate reasons.[28] Although these debates have claimed to address the effects of inequality broadly understood, they have typically focused on

24. Archon Fung, "Survey Article: Recipes for Public Spheres: Eight Institutional Design Choices and Their Consequences," *Journal of Political Philosophy* 11, no. 3 (2003): 338–67.

25. Jack Knight and James Johnson, "What Sort of Equality Does Deliberative Democracy Require?," in *Deliberative Democracy: Essays on Reason and Politics*, ed. James Bohman and William Rehg (Cambridge, MA: MIT Press, 1997), 279–321; Lynn Sanders, "Against Deliberation," *Political Theory* 25, no. 3 (1997): 347–76; Young, "Activist Challenges to Deliberative Democracy."

26. Quote from Jürgen Habermas, *Between Facts and Norms*, trans. William Rehg (Cambridge, MA: MIT Press, 1996), 306. His critics on this point include Joshua Cohen, "Deliberation and Democratic Legitimacy," in *The Good Polity: Normative Analysis of the State*, ed. Alan Hamlin and Philip Pettit (Oxford: Blackwell, 1989), 67–92; Amy Gutmann and Dennis Thompson, *Why Deliberative Democracy?* (Princeton, NJ: Princeton University Press, 2004).

27. Fung, *Empowered Participation*; Jane Mansbridge, James Bohman, Simone Chambers, David Estlund, Andreas Føllesdal, Archon Fung, Cristina Lafont, Bernard Manin, and José Luis Martínez, "The Place of Self-Interest and the Role of Power in Deliberative Democracy," *Journal of Political Philosophy* 18, no. 1 (2010): 64–100.

28. For a critical evaluation of these efforts, see Samuel Bagg, "Can Deliberation Neutralise Power?," *European Journal of Political Theory* 17, no. 3 (2018): 257–79; Simone Chambers,

inequalities in wealth, class, gender, and race, and paid little attention to the specific difficulties caused by inequalities in knowledge and expertise.

The neglect in the literature is understandable since it is not clear at first whether asymmetry in knowledge among deliberators is similar to inequalities of wealth or gender in that it should be irrelevant to the outcome of deliberation, or it is more similar to and perhaps correlated with the quality of arguments—that is, precisely what the outcome of deliberation is meant to track. To determine whether inequalities in knowledge pose a problem for the deliberative ideal, it is important to be clear about what kind of equality is required for deliberation. As Jack Knight and James Johnson have pointed out, deliberation cannot aspire to equality of influence over the outcome; that would defeat the purpose.[29] Deliberation aims to discriminate between competing ideas based on their quality and justifiability with the goal of producing better arguments and conclusions. Good and bad ideas cannot be treated alike, which means that some people will—and should—have more influence than others. The relevant conception of equality for deliberation must therefore focus on procedure rather than outcome. Habermas has argued that ideal deliberation requires "a symmetrical distribution of the opportunities for all possible participants to choose and perform speech acts."[30] It also requires that differences in influence over the final outcome of deliberation be insensitive to inequalities in factors such as resources, power, gender, or race, which should be irrelevant from the perspective of the better argument. Inequalities in influence should be due to the differences in the quality of reasons and arguments.

To put it more analytically, we could say that the equality condition for deliberation has two components: the speakers' opportunity to speak, and the listeners' uptake of what is spoken. Expert-layperson deliberation is likely to be problematic on both these dimensions. Laypeople will not have the same level of knowledge as experts and will not be able to express their knowledge in expert vocabulary. When political issues with a significant scientific component are under consideration, laypeople will find it difficult to outwit

---

"Rhetoric and the Public Sphere: Has Deliberative Democracy Abandoned Mass Democracy?," *Political Theory* 37, no. 3 (2009): 323–50.

29. Knight and Johnson, "What Sort of Equality Does Deliberative Democracy Require?," 295.

30. Jürgen Habermas, *On the Pragmatics of Social Interaction*, trans. Barbara Fultner (Cambridge, MA: MIT Press, 2001), 98.

experts. Studies of continental mixed juries of professional judges and lay citizens—a long-standing site of expert-layperson deliberation—support this view: mixed tribunals were found to reduce laypeople to "nodders" with a tendency to follow the lead of the professional judges.[31]

This is not to suggest that laypeople will have nothing to contribute. Even if they lack the knowledge to counter expert claims, they can nonetheless evaluate and judge the reasons and evidence provided by experts. They can ask questions, look for signs of bias, and evaluate the persuasiveness of the arguments presented. Although these contributions could be part of deliberation, it is important to distinguish between deliberation aimed at mutual justification and an exchange aimed at criticism of competing views. If an interaction is more likely to involve a critical exchange between those who are unequal in knowledge, such as between experts and nonexperts, rather than a mutual give and take of reasons and justifications between presumed equals, then it is necessary to rethink the institutional design appropriate for the interaction. The goal should be to design institutions that will facilitate the critical scrutiny of views put forward under circumstances of asymmetrical knowledge, and expose as much as possible the limitations, errors, and uncertainty of different views so that those who do not know are given the best possible chance for judgment.

Deliberation may well open up different views to criticism and expose their weaknesses, but it is not guaranteed to do so. Whether it ends up doing so depends on the availability of different perspectives.[32] On complex technical subjects, opposing views are most likely to come from other experts. Experts are best positioned to criticize each other and expose the weaknesses of each other's claims. Moreover, speakers in a political context may suppress information or present it selectively for strategic reasons. This may be intended to manipulate or simply to ensure that the audience forms what the speaker thinks are the correct beliefs. Experts might worry, for instance, that providing the full information may mislead nonexperts to take the findings to be less certain than they are. These are common features of persuasion and not unique to expertise. What makes them intractable in expert contexts is that a nonexpert audience may have no way of discovering the suppressed information or contesting the

31. Valerie Hans, "Jury Systems around the World," *Annual Review of Law and Social Science* 4 (2008): 276.

32. Bernard Manin, "Political Deliberation and the Adversarial Principle," *Daedalus* 146, no. 3 (Summer 2017): 28–38.

selective emphasis. Nonexperts are relatively helpless in contexts of expertise, and this helplessness is mutual knowledge.[33] In deliberation among epistemic equals, if participants suppress reasons or presents the facts selectively, there is a good chance that others who disagree will challenge them. In expert cases, only other experts may be in a position to detect the suppressed information or contest the selective emphasis. Nonexperts will be better able to evaluate expert views only if they receive information about alternative expert claims and observe experts criticize each other. This cannot be guaranteed to happen in a deliberative setting; it must be actively promoted.

The difficulty of submitting expert claims to public scrutiny has led some theorists to give up on the possibility of meaningful interaction between experts and laypeople on technical matters. Philosopher John O'Neill has argued that experts giving reasons should be seen as a rhetorical device to signal credibility and trustworthiness.[34] The evidence and arguments cited by the expert should not be interpreted as aiming to persuade the listener of the validity of a conclusion (since the listener could not evaluate this) but instead serve as a demonstration of the speaker's intention to share power with the audience. Reasoned argument thus serves as testimony to the speaker's good character. The work of persuasion is done by demonstrated trustworthiness rather than by the content of what is said. The problem is that neither the demonstrated ability to make reasoned arguments nor openness, trustworthiness, or willingness to share power closely tracks scientific knowledge. Moreover, being restricted to second-order assessments limits nonexperts' power vis-à-vis experts in deliberative settings. This model is designed to enable experts to gain the trust of nonexperts as opposed to helping nonexperts to scrutinize expert claims.

There is no easy solution to the difficulties of deliberation between experts and laypeople under circumstances of unequal knowledge and authority. Still, it is preferable to think about institutional arrangements that would be better suited to allowing nonexperts to examine expert claims rather than tackling the problem only in terms of trust cultivation and an ethics of experts, or ignoring it entirely.[35]

33. Alexander Guerrero, "Living with Ignorance in a World of Experts," in *Perspectives on Ignorance from Moral and Social Philosophy*, ed. Rik Peels (New York: Routledge, 2017), 156–85.

34. John O'Neill, "The Rhetoric of Deliberation: Some Problems in Kantian Theories of Deliberative Democracy," *Res Publica* 8 (2002): 249–68.

35. On expertise and trust cultivation, see Michael Fuerstein, "Epistemic Trust and Liberal Justification," *Journal of Political Philosophy* 21, no. 2 (2013): 179–99.

A Proposal for a New Science Court

The main weakness of Kantrowitz's proposal for a science court was its unrealistic assumption that the facts of a scientific dispute could be separated from the values involved. I argued in chapters 2 and 3 that the scientific claims to be accepted as the basis of a policy decision cannot be determined without considering their meaning and significance in a context, the background assumptions and value judgments involved, and moral considerations about the consequences of action, given uncertainty and the possibility of error. Kantrowitz maintained that the judges should be scientists because of their superior ability to assess scientific evidence and the presumptive authority that this competence would lend to the decisions.[36] If the matter were a purely scientific one, these might be the only considerations. But given the difficulty of separating the facts and values, the authority of the judge cannot be justified purely on the grounds of competence. The science court should therefore be treated as a political institution, and the authority of the judge must be justified on democratic rather than scientific grounds. An institution for the democratic use of science in policy cannot have experts decide the facts without scrutiny.

The new science court that I propose would address a scientific policy question in the form in which it would face policy makers, evaluating the facts and values together. The court would be initiated by ordinary citizens, and its decision would advise political decision-making. The proceedings would involve competing experts making the case for different sides of a scientific question, followed by a period of cross-examination and interrogation by the jury. The jury, which would be made up of randomly selected citizens, would deliberate and deliver a decision on the question posed to the court. Let me say more about each of its institutional features in turn.

Initiation of the Court and Agenda Setting

The science court would examine policy questions with a significant component of scientific knowledge. Climate change, pandemics, environmental disputes, nuclear waste disposal, biotechnology, and vaccines are some current issues that could be addressed. Since the science court is intended to empower citizens to engage with expertise on their own terms, they should

---

36. Arthur Kantrowitz, "Controlling Technology Democratically," *American Scientist* 63 (1975): 505–9.

have the power to initiate the proceedings. The court would be initiated through a petition with a number of required signatures.[37] Anyone could act as petitioner; the required number of signatures would vary with the relevant local or national population. The petitioner would have control over the framing of the question, but the institutional features of the court would necessarily constrain the questions that could be examined. Adversarial arrangements are best suited to examining questions with a binary structure. The number of positions represented could be up to three or maybe four, but not more. This restricts the issues that could be addressed in a science court, but it also enforces simplicity, which is extremely valuable for facilitating public comprehension and participation. The court would be better suited for settling questions on which there are clearly defined rival positions and where at least part of the disagreement is over the science. Questions with a yes or no answer, or choice between a few policy proposals, would work best. The court would not be ideal for creative policy development; Fung's deliberative problem solving may be more appropriate where that is the goal.

Citizen groups are unlikely to have policy proposals of their own, so they could choose to have the court examine proposals already on a legislative agenda, recommended by scientists, or developed by NGOs or think tanks. The majority and minority opinions from a scientific advisory committee would be ideal material for examination. Even if the main ideas and options are derived from expert sources, it would be important for the framing of the question to go through the filter of citizens' judgment and selection rather than being determined by the organizers, as is usually the case. To make the framing task more straightforward, the initiators could opt for simple yes or no questions. Here are some examples of the sorts of questions that could be (or could have been) taken up in a science court:

- The government should impose a national lockdown to slow the spread of COVID-19 (yes/no)
- Wolves should be reintroduced to Yellowstone National Park (yes/no)
- Xenotransplantation should be legal (yes/no)

37. This is similar to the initiation of the Oregon Citizens' Initiative Review. See John Gastil, Robert C. Richards Jr., and Katherine R. Knobloch, "Vicarious Deliberation: How the Oregon Citizens' Initiative Review Influenced Deliberation in Mass Elections," *International Journal of Communication* 8, no. 1 (2014): 62–89.

- Anthrax vaccines are safe enough to be tested on children (yes/no)
- Hormone replacement therapy should be recommended to all women after menopause (yes/no)
- Large-scale geoengineering experiments should be banned (yes/no)
- Climate policy A, B, or C should be adopted to meet specified emissions target by a certain year

The framing of the question would be publicized for comments and suggestions before the collection of signatures begins. The petitioner would retain the right to revise the wording. The petitioner would also be responsible for selecting experts to make the case for at least one side of the issue. Scientists from several disciplines would most likely be involved in developing the case, along with nonscientists possessing relevant experiential knowledge on the issue. Recent scientific assessment reports might provide good material for public examination in a court, especially if they involve dissenting views. The court would benefit from a database that lists experts by domain of their expertise, credentials, experience, and track record to help with the selection.[38] This would facilitate public scrutiny of track records and potential conflicts of interest.

The professional staff organizing the science court would be responsible for finding experts to play devil's advocate, defending the view opposed to the petitioner's. The science court could not function without the organizers, but the organizers' role would be more restricted than in existing minipublics and circumscribed by the agenda set through a petition. To give an example, the farmers in the Minnesota power line case would initiate a science court to examine the claims about the health, safety, and environmental risks posed by the proposal. They would invite scientists to make the case for the risks posed by the power line, and the organizers would invite scientists willing to defend the view that the risks are not substantial enough for concern.

To ensure that a lack of resources does not pose an obstacle to citizens who want to bring an issue to a science court, government funds could be made available for the organization of the event, such as from the National Science Foundation's budget. Although citizen-initiated science courts should be the norm, the government or a group of scientists would also have the right to initiate. In these instances, the government or scientists would set the agenda

38. For a similar proposal for a database of experts, see Guerrero, "Living with Ignorance in a World of Experts," 180.

and select experts to make one side of the case, and the science court organizers would put together a team of scientists to make the strongest case against
it. In certain policy areas, it might be desirable to mandate the organization of
a science court before decisions are made.

Adversarial Proceedings

Adversarial proceedings are ideal for examining the grounds of competing
claims and revealing questionable assumptions and errors through confronta-
tion with opposing views. The underlying principle is that the validity of truth
claims can best be tested if they are tried out against the strongest opposing
arguments. Of course, ordinary discussion can bring rival perspectives into
conflict and produce the same results too. Mill held this up as one of the chief
benefits of free discussion in *On Liberty*. Yet it cannot be assumed that freedom
of discussion alone would be sufficient to ensure that opposing views would
emerge and a critical examination of opinions would take place. Nor could the
creation of deliberative bodies ensure this. Critical exchange could happen in
deliberation, but it is not guaranteed. The crucial difference between adversarial and nonadversarial arrangements is that the former forces a confronta-
tion between opposing views by design, while the latter does not.[39]

   Alvin Goldman's account of how a novice can choose between two competing experts supports the epistemic value of adversarial arrangements. Goldman argues that a novice may be able to evaluate experts in a setting where
"two experts might engage in a full-scale debate that N[ovice] witnesses. . . .
Each expert might there present fairly developed arguments in support of her
view and against that of her opponent."[40] The novice who observes this performance could determine which expert has the better reason to believe their
conclusion based on dialectical superiority—that is, the ability to justify one's
position, answer challenges, come up with further evidence, and provide effective rebuttals. The novice's judgment in this case is indirect, and consists of
an inference from the effectiveness of arguments and challenges to the conclusion that one of the experts has better reasons in support of their claim. While
this is different than the expert's own justification, which derives directly from
the support relation between the evidence and claim, it nonetheless results

39. Manin, "Political Deliberation and the Adversarial Principle."
   40. Alvin Goldman, "Experts: Which Ones Should You Trust?," *Philosophy and Phenomenological Research* 63, no. 1 (2001): 93.

from serious first-order engagement with the content of expert arguments about the evidence. It should therefore not be confused with unreliable second-order assessments about rhetorical polish, smoothness, demeanor, or appearance. Melissa Lane elaborates on Goldman's argument by explaining that this indirect assessment requires the skill, virtue, and habit of good judgment, which consists in the ability to engage in certain kinds of critical reasoning and inference, and detect these patterns in others. This skill can be possessed and cultivated by expert and nonexpert alike.[41]

The adversary process in a science court could work in two different ways. The first is to invite only scientists to defend different positions and cross-examine each other. This would be appropriate for simpler issues, with less disagreement over the values at stake. "The water supply should be fluoridated" would be an example. The alternative is a structure more similar to a traditional court, where those for and against a proposal would present a case, call in experts, and cross-examine experts called by the other side. Stakeholders, interest groups, activists, or NGOs could be involved in developing cases and inviting experts. This structure would be more appropriate in more complicated issues and where there are rival stakeholders invested in building expertise on the matter. "GMOs should be labeled" would be more suited to this structure.

Citizen Jury

My proposal's most significant departure from Kantrowitz's science court is in replacing scientist judges with a lay jury. The jury would be made up of ordinary citizens selected through random sampling from the relevant national or local jurisdiction, excluding those who initiate the court. I defend this primarily on democratic grounds. The jury in this context should be considered primarily a political body rather than a scientific or judicial one. Arguments in favor of this proposal will be analogous to those for jury trials. I will

41. Melissa Lane, "When the Experts Are Uncertain: Scientific Knowledge and the Ethics of Democratic Judgment," *Episteme* 11, no. 1 (March 2014): 97–118. For more detailed discussions of how laypeople can adjudicate between experts, see Brown, "Expertise and Deliberative Democracy"; Moore, *Critical Elitism*, chapter 4. The legal literature on innovations for assisting jurors in making sense of expert testimony is also helpful. See, for example, Neil Vidmar, Shari Seidman Diamond, Mary R. Rose, and René Stemple Ellis, "Juror Discussions during Civil Trials: Studying an Arizona Innovation," *Arizona Law Review* 45 (2003): 1–83.

list three justifications for jury trials and show how each one can be transposed to the science court.

The first and most important justification for juries is that they are an expression of popular sovereignty—of the people's control over their agents in government.[42] Juries allow citizens to check the power of the state, judges, and (in civil trials) corporations. The use of a citizen jury in a science court would be justified on the same basis. Ordinary citizens would have the opportunity to check the claims of experts and participate in the shaping of policy on scientific issues.

A second justification for jury trials is that juries inject local knowledge and community values into legal procedures.[43] This justification can also be transferred to the science court. Local knowledge would be particularly relevant on issues specific to areas or groups, such as in environmental disputes or medical research on specific patient populations. In these cases, nonexperts might possess experiential or observational knowledge that experts would have difficulty acquiring. The argument for community values follows a different logic. The point is not that ordinary citizens would bring predefined community values that they would somehow know better than experts. They would instead be asked to examine the values and purposes driving scientific claims in light of their personal values and perspectives. What would make citizen jurors preferable to a panel of experts is that the former would be representative of public opinion in a descriptive sense because of their method of selection from the population. Ordinary citizens would bring different purposes and priorities than scientist judges, who would be more likely to share the professional concerns of the testifying experts. Moreover, the jurors might be more diverse in terms of background and opinions, whereas experts tend to be a relatively homogeneous group.[44]

A third justification for the jury system is that it gives ordinary citizens the opportunity to understand the workings of the legal system and increases their

42. Akhil Amar, "The Central Meaning of Republican Government: Popular Sovereignty, Majority Rule, and the Denominator Problem," *University of Colorado Law Review* 65 (1993): 749–87; Akhil Amar, "Reinventing Juries: Ten Suggested Reforms." *UC Davis Law Review* 28 (1995): 1169–95.

43. Jeffrey Abramson, *We, the Jury: The Jury System and the Ideal of Democracy.* New York: Basic Books, 1994.

44. Hélène Landemore, *Democratic Reason: Politics, Collective Intelligence, and the Rule of the Many* (Princeton, NJ: Princeton University Press, 2008); Scott Page, *Diversity and Complexity* (Princeton, NJ: Princeton University Press, 2010).

assurance that it is functioning properly (if indeed it is). Making the system intelligible to citizens allows them to feel that the forces that determine their fate are not beyond their control. This point is crucial for scientific issues. The complexity of science and technology makes it particularly difficult for citizens to understand and control policy on these issues, which can lead to a feeling of alienation. An institution that allows citizens to get directly involved on issues requiring expertise and listen to firsthand accounts from the experts would serve an important demystifying purpose. This would make citizens feel less removed and more powerful in technical areas of policy.

One objection that I anticipate to the use of a citizen jury is that it amounts to putting scientific truth to a popular vote. The objection misses the distinction between the truth of a scientific theory and its acceptance in a science court. The decision about which facts to accept in this context is not a decision about what is true; it is a decision about what can be relied on for the purposes of a particular policy given that there is disagreement about what the truth is, and the science is uncertain and incomplete. The science court's evaluation of expert claims and its final judgment are oriented toward what to do rather than what to believe. This requires practical judgment rather than theoretical judgment. There is an important difference in this respect between the decision of a science court and Goldman's setup of a layperson asked to identify which of two experts is the correct one. A decision about what to do depends partly on what is taken to be true, but it also depends on practical considerations about whether the evidence on either side is sufficient for action, how complete the available scientific knowledge is, how bad it would be to make different kinds of mistakes, and which policies would be better to pursue under uncertainty or ignorance. The science court has no claim to settling theoretical controversies for the scientific community.

Admissibility of Expert Evidence

Expert arguments and juror questions must be relevant to the purpose of determining the truth and settling the policy question under consideration. The *Federal Rules of Evidence* offers the following test for judging the relevance of a piece of evidence: it "is relevant if it has any tendency to make a fact more or less probable than it would be without the evidence; and the fact is of consequence." Moreover, "the court may exclude relevant evidence if its probative value is substantially outweighed by a danger of one or more of the following: unfair prejudice, confusing the issues, misleading the jury, undue delay, wasting time, or needlessly presenting cumulative evidence."[45] Both of these rules are relevant for the science court, with the possible exception of the prejudice restriction. The institution would therefore benefit from having a moderator to judge the relevance of expert evidence and juror questions, and make sure that the court does not veer off course.

This raises a more fundamental question about admissibility standards for expert testimony. Should all purportedly scientific views be accepted in a science court, or should there be a filtering process to determine which views to admit? I anticipate the worry that the science court could become a platform that spreads and even legitimizes "unreasonable" scientific views unless it has a reliable filtering mechanism. Any response to this concern must start from an account of what constitutes an unreasonable scientific view in this context and why.

This question is easier to answer in the abstract. Following Inmaculada de Melo-Martín and Kristen Intemann, I propose two criteria for identifying views that would not be productive to examine in a science court.[46] The first criterion is bad faith, which describes claims intended to mislead, confuse, or deceive others. Views based on fabricated data, manufactured uncertainty, misrepresented evidence, and selectively reported findings would violate this criterion. This would disqualify the scientific claims that corporations have put forward to create doubt on issues such as climate change and the health effects of tobacco.[47] The second criterion is the failure to play by the rules of scientific exchange and critical debate, by not responding to objections, refusing to justify one's view, avoiding cases that pose difficulties for one's theory, not changing one's position in response to new evidence, and so on.

The initiators of the science court would be responsible for applying these two criteria and selecting views that exemplify good faith scientific disagreement. The real difficulty arises at this point: What if those who initiate the court fail to apply the criteria reliably and end up selecting unreasonable views? Would it not be better for a body of experts to act as gatekeepers to filter out

45. *Federal Rules of Evidence* (Grand Rapids, MI: Michigan Legal Publishing Ltd., 2021), 401, 403.

46. Inmaculada de Melo-Martín and Kristen Intemann, *The Fight against Doubt: How to Bridge the Gap between Scientists and the Public* (New York: Oxford University Press, 2018).

47. Naomi Oreskes and Eric Conway, *Merchants of Doubt: How a Handful of Scientists Obscured the Truth on Issues from Tobacco Smoke to Global Warming* (New York: Bloomsbury Press, 2010).

the unreasonable views? After all, experts have a better understanding of the quality of different scientific claims and the ability of other experts to respond adequately to objections. This is a basic premise of the science court.

The problem is that these two criteria are not determinate enough to be applied reliably and uncontroversially, whether by experts or nonexperts. De Melo-Martín and Intemann analyze the most widely acknowledged cases of science denial—on climate change, vaccine safety, and evolution—and conclude that the application of these criteria turn out to be ambiguous and subjective in each instance. Let me summarize their main points: the motivations of researchers are difficult to establish, bad motivations may still yield epistemically fruitful dissent, and many strategies of sophisticated bad faith dissenters are indistinguishable from good faith efforts to highlight uncertainties. Having a conflict of interest or value-based sympathy for a scientific position is not enough to disqualify one's arguments and evidence. What constitutes a willingness to play by the rules of scientific exchange is also subjective. No one can be expected to accept all criticism, and those who deeply disagree are likely to find each other's responses to criticism inadequate. Finally, any requirement of shared standards is open to interpretation. It is not clear how many shared assumptions are sufficient and how much agreement on the interpretation of each standard is required. Stringent interpretations of rules will exclude genuine and productive dissent, while relaxed ones will allow views widely believed to be unreasonable.

This does not mean that these criteria are not useful; to the contrary, they may be the most useful criteria available. Citizens and policy makers should apply them to the best of their judgment. The point is rather that their application is not just a matter of scientific competence, and remains prone to epistemic and political bias. Whoever has the authority to filter views for the science court will thus be doing a version of the court's work of examining and judging the quality, completeness, and bias of different scientific views under conditions of disagreement over facts, values, and background assumptions. Entrusting this task to a group of experts would undermine the rationale for the science court, which is to allow nonexperts to make judgments on questions with this structure. It would also generate distrust of the institution and give the impression that the process is steered by experts.

This might raise the question of whether the democratic gains from the science court would be worth the risk of giving a platform to unreasonable views, which cannot be eliminated fully. Unfortunately, it is impossible to settle a weighing problem of this sort without an all-things-told evaluation of the consequences of the court. This would require empirical evidence that cannot be obtained without testing the proposal. Extrapolation from existing institutions such as courts or minipublics is inadequate because of crucial differences in design, and because science court outcomes are likely to be highly contextual.

Since I have already described the possible democratic benefits, I will simply say something in response to the worry about giving a platform to unreasonable views. If a large number of citizens request to have certain scientific claims examined in a science court, it means that these views already enjoy a large platform. Scientific dissents are often funded by corporations, have direct access to politicians through well-paid lobbying efforts, receive publicity from partisan media organizations and celebrities, and may even gain legitimation through publication in peer-reviewed journals, as in the measles, mumps, and rubella vaccine case. Some dissenters organize their own conferences, create new journals, and found think tanks to spread their views. While the scientific community has its own gatekeeping mechanisms, no comparable ones exist to prevent the spread of these perspectives in the public sphere. The inclusion of these views in a science court is therefore unlikely to make a big difference to the size of the audience that these views get. Ultimately, the decision to include a view in a science court shows that a number of people find it credible, relevant, and advanced in good faith. Without respectable public platforms where the falsehood and bad faith of these claims can be exposed, even more people might begin to find these views credible. This danger would be especially great if segments of the population encounter defenses of these views over and over again through social media and networks without sufficient opportunities to hear criticisms, takedowns, and alternative stances.

Jury Deliberation and Decision-making

After the expert cross-examination period, the jurors would deliberate and take a vote on the policy question as it was originally posed to the court. The deliberation would evaluate the different claims and evidence presented by the experts, focusing on both the scientific aspects and value judgments required to reach a decision. For instance, in a discussion about whether GMOs should be banned, the jury would consider the effects on local biodiversity, sustainability, and food security, reflect on the proper conception of safety in this context, weigh the potential benefits to some people over the potential harms to others, and determine whether the evidence presented is sufficient for a ban given the consequences of error.

Juror deliberations would be conducted in secret to allow jurors to express their opinions more candidly. Secrecy could increase the quality of deliberation by freeing the jurors from worries about how the public or media might respond to their views.[48] Jurors might be more willing to compromise, admit their mistakes, and change their minds in secret deliberations. Since the topics discussed in a science court would be unfamiliar to most jurors, secrecy may also be important for making jurors feel comfortable enough to talk about the points they may not have understood, admit their mistakes, ask for clarifications, and venture their own opinions. Furthermore, secrecy could insulate deliberators from possible pressures from stakeholders and interest groups. The issues taken up in a science court will often be high stakes and politicized, involving many vested interests. One of the advantages of this institution over regulatory agencies and elected representatives is that it can be insulated from capture since the jurors are selected randomly and serve only once. But the realization of this benefit depends on whether jurors can be shielded from external efforts to delegitimize the process. Secrecy can be instrumental for this. There is a trade-off between the democratic benefits of secrecy and publicity. Since the priority of this institution is to make experts accountable to the public and facilitate citizen participation on scientific issues, it would be preferable to make the trade-off in favor of secrecy.[49] The rest of the proceedings would be public, and everyone would have the opportunity to hear and evaluate the evidence provided by experts.

The jury deliberations would conclude with a vote on the question under consideration. The vote could affirm one of the competing positions, endorse an intermediate position, or state that different sides were persuasive on different subparts of the question, if applicable.[50] The jury could also point out that the evidence is too uncertain or incomplete to act on and suggest that more research is necessary. Since most science is uncertain and incomplete, this judgment would indicate that the jury judges the status quo to be

48. Amy Gutmann and Dennis Thompson, *Democracy and Disagreement* (Cambridge, MA: Harvard University Press, 1996); Simone Chambers, "Behind Closed Doors: Publicity, Secrecy, and the Quality of Deliberation," *Journal of Political Philosophy* 12, no. 4 (2004): 389–410.

49. On how the unaccountability of the demos was justified in ancient Athenian democracy, see Matthew Landauer, "The *Idiōtēs* and the Tyrant: Two Faces of Unaccountability in Democratic Athens," *Political Theory* 42, no. 2 (April 2012): 139–66.

50. This would be similar to civil juries allocating percentage fault between the two parties, as opposed to a declaration of guilt or innocence in a criminal trial.

preferable given the potential consequences of action and inaction under uncertainty.

How the jury should present its judgment poses a bit of a dilemma. On the one hand, it is unrealistic to expect a jury of laypeople to provide a sophisticated justification of its decision on a complex scientific matter. The adversarial court setup is meant to facilitate the participation of ordinary citizens on issues where they might have difficulty making arguments. To demand detailed and scientific reasoning in support of a judgment would simply re-create the original challenges to participation. The responsibility of the jury should be to reflect on the evidence presented, not to make its own case in support of a position. This is standard practice in jury trials, where juries are not expected to offer any reasons for their decisions on matters literally of life and death. An important difference between a legal trial and science court, however, is that the latter is concerned with matters of policy, which will ultimately be binding on others. Since the science court jury is composed of unelected citizens, it would not be right for it to have the authority to enact policies or make binding laws, regardless of whether the jury gives reasons for its decision. The jury's decisions should advise the legislature, president, or executive agencies.

The limited scope of the jury's direct decision-making authority is a reason not to be too worried about juror unaccountability. Still, an advisory body's failure to give reasons and justifications may diminish its influence over policy makers. A brief report summarizing the arguments that were most persuasive, the background assumptions that were found problematic, and the harms and benefits that were most significant for the jury's decision might be helpful for policy and increase the likelihood that the science court decision is accepted. A science journalist or graduate student might serve as secretary to the jury to help with the composition of a report without participating in the deliberation.

The Role of the Science Court in the Political System

I now want to clarify the democratic role of the science court, focusing on its relationship to both formal policy processes and informal political discussion. The science court is intended to have direct advisory influence on policy makers. In this respect, its official status is similar to that of a scientific advisory committee. Yet the authority of the science court is not derived from a superior claim to scientific competence but instead from a democratic claim to descriptively representing the public, given the random selection of jury members. The initial similarity between citizen jurors and the population as a whole suggests that nonparticipants would have reached a similar conclusion had they participated in the process. Deference to the science court's decision would be justified on the basis that it is a reliable indicator of the attitude that other citizens would have developed if they had likewise engaged directly with expert views.[51] Philip Pettit has coined the term "indicative representation" to describe this form of representation (to be contrasted with the "responsive representation" of elected officials) and has argued that it presents an alternative source of democratic credentials—one that relies on the idea that an accurate cross section of the community can stand in for the whole.[52] This, of course, is the idea behind the jury trial too.

The main advantage of the science court is that it can make up for the democratic problems with deference to unexamined expert opinion while overcoming some of the epistemic shortcomings of uninformed public opinion. The last two chapters made the case that scientific advice should not be accepted without scrutiny, and this scrutiny should be democratic. Scientific advisory committees alone cannot be given full authority on issues requiring expertise, and representatives cannot be fully trusted to scrutinize expert advice, especially when their private interests go against doing so. Once the option of leaving issues of expertise entirely to experts or elites is rejected on democratic grounds, the challenge is to ensure that citizens can participate most effectively on scientific issues and have the best opportunity for holding experts accountable. The science court is designed to facilitate this.

The science court's legitimacy is thus derived on democratic grounds first and foremost, but with the added benefit of institutional rules designed to create the most favorable epistemic conditions for scrutiny under conditions of asymmetrical knowledge. Face-to-face interaction with experts, adversarial rules, and the time spent considering the case are intended to improve the jurors' understanding and scrutiny of complex technical knowledge. Still, these are ultimately secondary to the court's democratic function. The epistemic credentials of the court could perhaps be improved by adding a number of experts to the jury or filtering jurors according to scientific background, but this would undermine the court's democratic credentials. The relative weights

51. Moore, *Critical Elitism.*

52. Philip Pettit, "Representation, Responsive and Indicative," *Constellations* 17, no. 3 (2010): 426–34.

attached to democratic and epistemic advantages in a science court are hence diametrically opposed to those in expert committees, which are legitimated primarily on epistemic grounds, with some efforts at improving democratic representation through the inclusion of diverse viewpoints. Since the kinds of questions that would be taken up in a science court combine practical and value-based considerations with the scientific, legitimation on primarily democratic grounds is more important. Procedurally, deference to a science court therefore has more in common with deference to elected representatives than to an expert committee; both are legitimated on the basis of a claim to democratic representation.

The science court would also reduce the inherent tension of the role of experts in the policy-making process—namely that policy making would lose democratic legitimacy if experts acquired a lot of power over representatives and would lose the benefits of expertise if experts had no influence at all. If a citizen jury scrutinized expertise, then it would be both desirable and legitimate for expertise to acquire more authority to shape decisions. Although the science court would not have authority to enact its decisions, various measures could be taken to increase its political influence and make it difficult for decision makers to ignore its outcome. Citizen groups could seek an advance public commitment from policy makers that they would follow the court's decision. Officials could be required to initiate a second science court on the same question if they rejected the recommendations of a first one. Only after a second court delivered a decision would they be allowed to disregard the recommendations.[53] The court could acquire influence with policy makers through wider public support too. This is not difficult to imagine if it were a televised event on a national issue such as climate change. The arguments and counterarguments could shape public opinion, and the court might gain the trust of the public if it is widely perceived to be a fair, democratic, and well-run process.[54] In fact, it has been argued that a crucial role for minipublics would be to serve as trusted proxies under conditions of widespread distrust in institutions that have traditionally played this part and where high cognitive

53. For a similar proposal in a different context, see Bruce E. Cain, "Redistricting Commissions: A Better Political Buffer?," *Yale Law Journal* 121 (2012): 1808–44.

54. Dennis Thompson, "Who Should Govern Who Governs? The Role of Citizens in Reforming the Electoral System," in *Designing Deliberative Democracy: The British Columbia Citizens' Assembly*, ed. Mark E. Warren and Hilary Pearse (Cambridge: Cambridge University Press, 2008), 20–49.

demands make it difficult for citizens to evaluate information on their own.[55] If public trust in science courts is widely acknowledged, its proposals would come to bear more weight in the policy-making process.

Since the science court would not be the only voice in the public sphere, it is important to consider how the court's decisions would relate to broader currents of public opinion. Following Cristina Lafont, we can analyze this question by focusing on three possibilities: public opinion on the issue is not yet settled; public opinion is settled and the science court aligns with the majority view; and public opinion is settled and the science court aligns with a minority view.[56]

Many issues taken up in a science court would fit the first case, at least initially. New scientific and technological findings often set the agenda, and are handled through elite channels without public discussion. For instance, although scientists, activists, and politicians were aware of the problems posed by climate change and debated possible responses to it in the 1980s, the issue did not reach mainstream public debates until the 1990s. A crucial function of the science court would be to make scientific issues visible and salient from the early stages, stimulating public debate and contributing to the framing of the issue for subsequent political discussion. The court would create awareness and interest in cases where the public might be unaware or indifferent— frequently until it is too late. Of course, there must be a certain number of citizens sufficiently concerned with an issue to initiate the court. These could be activists or those directly affected by a problem. But once the court is initiated, it would spread awareness to the rest of the population. The initiators' framing of the question would itself be a significant contribution to public debate since it would provide an alternative to the frames that scientists, the government, or industry might offer. The court could reclaim for ordinary citizens some of the framing power usually exercised by elites.[57]

In other cases, the court would take up issues on which public opinion is already formed, and rival positions are clearly defined. The court would then be expected to resolve an existing dispute. In these instances, the court

55. Mark E. Warren and John Gastil, "Can Deliberative Minipublics Address the Cognitive Challenges of Democratic Citizenship?," *Journal of Politics* 77, no. 2 (2015): 562–74.

56. Cristina Lafont, "Can Democracy Be Deliberative and Participatory? The Democratic Case for Political Uses of Mini-Publics," *Daedalus* 146, no. 3 (2017): 85–105.

57. Simone Chambers, "Balancing Epistemic Quality and Equal Participation in a System Approach to Deliberative Democracy," *Social Epistemology* 31, no. 3 (2017): 266–76.

decision would either align with the majority position or a minority one. If it aligns with the majority, it would reinforce this position and increase pressure on politicians to act on it. If this position goes against the preferences of politicians or special interests, the court might effectively reveal that they are merely a self-interested minority. The court decision would also undermine elite attempts to dismiss raw public opinion as ill informed or confused. The authority of the conclusion would increase if a group of citizens who had spent days or weeks examining the issue reached the same conclusion as those who had most likely thought less about it. Moreover, the court would supply public debates with additional information, arguments, and criticisms derived from scientists' arguments and cross-examination. This would increase the quality of public debate and provide fresh materials for defending the majority position.

The last possibility is that the court rules against the prevailing opinion and aligns with a minority position. In this case, the court decision would challenge the majority and push citizens to rethink their views. It would become a dissenting voice in the public sphere, fulfilling the Millian function of challenging widely held opinions, and forcing others to either change their positions or find new, stronger defenses for them. This would invigorate other dissenting voices by adding authority and accessibly presented arguments to their case. After all, it is one thing for a minority to argue against a majority, and another for one kind of democratic majority—a representative group with a claim to having examined a range of competing expert views for several days—to argue against another.[58] It cannot be guaranteed that the court will change most minds, but evidence from existing minipublics shows that people do change their minds in response to the recommendations of a properly conducted citizen panel, especially if its rules and procedures have become familiar and trusted over time.[59] Even if it does not succeed in changing minds, it would allow citizens to enjoy the rational assurance of being right in their views.[60]

If public opinion is ultimately unswayed by the science court, then decision makers would simply have to consider the court's verdict alongside other views. It would be undemocratic to insist that the court decisions must necessarily trump public opinion; Lafont is right to point out that democracy cannot leave

58. Lafont, "Can Democracy Be Deliberative and Participatory?"

59. Gastil, Richards, and Knobloch, "Vicarious Deliberation."

60. Nancy L. Rosenblum, On the Side of the Angels: An Appreciation of Parties and Partisanship (Princeton, NJ: Princeton University Press, 2008).

the people behind. Nevertheless, the science court possesses a special combination of democratic and epistemic features, distinguishing it from expert committees and public opinion alike. This provides a reason to give its recommendations special weight and attention. Representatives must weigh the science court verdict against the strength and number of other perspectives in deciding what to do. This is no different than situations in which expert opinion differs from public opinion or public opinion itself is highly divided. The decision will rest on the judgment of the decision maker.

The Competence Objection

I anticipate the objection that this proposal overestimates citizens' competence. Would nonexperts be able to understand and question expert claims? Would such an institution not result in arbitrary decisions, divorced from scientific reality? There are two ways to respond to this. The first is to argue that this proposal would be an improvement over alternatives, given normative democratic constraints on the alternatives that would be acceptable. The second is to give reasons to believe that citizens do in fact have the capacity to perform this task.

The objection can be broken down into two parts. First, most citizens do not have the competence to examine expert claims. Second, it is irrational for them to invest time trying to acquire the knowledge and skill that would allow them to evaluate complex technical issues since their likelihood of influencing the outcome is slim, and time is scarce.[61] The science court addresses the worry about rational ignorance by modifying two of the key variables that make information gathering irrational: time and the small likelihood of influencing the outcome. The proposal makes it mandatory for the selected jurors to attend hearings. This forces the participants to dedicate a significant amount of time to listening to experts; they effectively form a captive audience. Moreover, the requirement that the jury must reach a decision that will directly advise the policy process changes the stakes for the participants; since this decision will have more influence over policy than the opinion of an ordinary person sitting at home, it is rational for jurors to invest mental energy to understanding the scientific components of the problem.

61. Russell Hardin, "Street-Level Epistemology and Democratic Participation," *Journal of Political Philosophy* 10, no. 2 (2002): 212–29; Ilya Somin, "Voter Ignorance and the Democratic Ideal," *Critical Review* 12, no. 4 (1998): 413–58.

As for the competence objection, we can test it by studying the evidence from jury trials, particularly in civil cases, where juries have to evaluate complex scientific testimony from a wide range of experts before reaching a verdict. The charge that juries are confused by scientific testimony and reach arbitrary verdicts based on the emotional appeals of lawyers and expert witnesses is widespread.[62] But empirical research indicates that juries in fact perform their task well; they evaluate experts based on the merits of their testimony rather than on their likability or credentials, and understand the purpose and effects of the adversary process.[63] Of course, researchers assessing the competence of jurors face the difficulty of positing an external standard of correctness. Since trial verdicts are not simply judgments about the truth of scientific claims, and the truth of competing scientific claims is contested among experts, finding an evaluative standard poses a significant, perhaps insurmountable methodological limit to these studies. Still, the existing studies have used a variety of different methods, which together present a consistent picture of good performance. These methods include asking the presiding judge what verdict they would have given and comparing it to the jury's verdict, asking external and independent experts to evaluate the competing expert testimony presented at the trial, interviewing jurors about their decision-making process to understand how much they engaged with the content of expert testimony, and observing actual or mock jury deliberations.

Let me briefly summarize the findings in the literature. Harry Kalven and Hans Zeisel's well-known study of jury trials in the 1950s found that in 80 percent of the six thousand cases studied, the presiding judges said that they would have given the same verdict as the jury. In the 20 percent where the judges disagreed with the juries, the disagreement was found to be unrelated to the complexity of the case.[64] More recent studies, motivated by the

62. See, among many others, Stephen Daniels and Joanne Martin, *Civil Juries and the Politics of Reform* (Evanston, IL: Northwestern University Press, 1996); Marcia Angell, *Science on Trial* (New York: W. W. Norton and Company, 1996).

63. Neil Vidmar, "The Performance of the American Civil Jury," *Arizona Law Review* 40, (1989): 849–901; Larry Heuer and Steven Penrod, "Trial Complexity: A Field Investigation of Its Meaning and Effects," *Law and Human Behavior* 18 (1994): 29–51; Neil Vidmar and Shari Seidman Diamond, "Juries and Expert Evidence," *Brooklyn Law Review* 66 (2001): 1121–81; Sanja K. Ivkovic and Valerie P. Hans, "Jurors' Evaluations of Expert Testimony: Judging the Messenger and the Message," *Law and Social Inquiry* 28 (2003): 441–82.

64. Harry Kalven and Hans Zeisel, *The American Jury* (Boston: Little, Brown and Company, 1966).

possibility that a study from the 1950s would not capture the effects of in-creased complexity, followed the same methodology, and found similar results in both criminal and civil cases, with complexity remaining irrelevant for ex-plaining cases of disagreement.[65] Three studies asking physicians to examine jury decisions in cases of medical malpractice, and another study that closely scrutinized thirteen "complex" cases found no evidence of jury irrationality.[66] One study that interviewed jurors in five cases involving scientific testimony found that a significant number of the jurors could articulate the main scien-tific issues and understood the basic points made by competing experts.[67] Another study interviewed fifty-five jurors who served on a range of cases including ones related to medical malpractice, workplace injury, product liabil-ity, asbestos, or motor vehicles, and found that the majority of the jurors could critically evaluate the testimony of experts and gave nuanced responses to questions about the evidence.[68]

Particularly interesting for my purposes are the findings from the Arizona Jury Project's videotaping of fifty jury trials, where jurors were allowed to question expert witnesses—a rare practice in most states due to concerns about compromising juror impartiality.[69] What is striking in this study is how detailed and probing some of the juror questions about science turn out to be.

65. Heuer and Penrod, "Trial Complexity"; Paula L. Hannaford, Valerie P. Hans, and G. Thomas Munsterman, "Permitting Jury Discussions during Trial: Impact of the Arizona Jury Reform," *Law and Human Behavior* 24 (2000): 359–82; Theodore Eisenberg, Paula L. Hannaford-Agor, Valerie P. Hans, Nicole L. Waters, and G. Thomas Munsterman, "Judge-Jury Agreement in Criminal Cases: A Partial Replication of Kalven and Zeisel's *The American Jury*," *Journal of Empirical Legal Studies* 2, no. 1 (2005): 171–207.

66. Mark I. Taragin, Laura R. Willett, Adam P. Wilczek, Richard Trout, and Jeffrey L. Carson, "The Influence of Standard of Care and Severity of Injury on the Resolution of Medi-cal Malpractice Claims," *Annals of Internal Medicine* 117, no. 9 (1992): 780–84; Frank A. Sloan, Penny B. Githens, Ellen Wright Clayton, Gerald B. Hickson, Douglas A. Gentile, and David F. Partlett, *Suing for Medical Malpractice* (Chicago: University of Chicago Press; 1993); Henry Farber and Michelle White, *Medical Malpractice: An Empirical Examination of the Litigation Process* (Cambridge, MA: National Bureau of Economic Research, 1990); Richard Lempert, "Civil Juries and Complex Cases: Taking Stock after Twelve Years," in *Verdict: Assessing the Civil Jury System*, ed. Robert Litan (Washington, DC: Brookings Institution, 1993).

67. Daniel W. Shuman, Elizabeth Whitaker, and Anthony Champagne, "An Empirical Ex-amination of the Use of Expert Witnesses in the Courts—Part II: A Three City Study," *Jurimet-rics* 34, no. 2 (Winter 1994): 193–208.

68. Ivkovic and Hans, "Jurors' Evaluations of Expert Testimony."

69. Vidmar et al., "Juror Discussion during Civil Trials."

Overall, there is empirical evidence against the claim that juries are incapable of understanding expert testimony and making sensible decisions in cases involving complex technical knowledge. The competence objection against the science court is not backed by long-standing evidence from the analogous institution of the civil jury trial.

One last thing I want to add on the question of competence is that it is important not to essentialize it by assuming that a person either has or does not have competence on an issue. Competence can be developed over time through practice and participation, or can be acquired if and when an issue becomes particularly salient—for instance, if one is called to jury duty on a scientific issue. Jurors can also be given training before the court proceedings on the key scientific concepts, evidentiary standards, and measures of uncertainty as well as the types of questions they could ask. Furthermore, public comprehension and competence can be improved through more effective communication strategies and more intelligible presentations of complex information. The literature on jury competence not only evaluates how well juries perform but also examines communicative strategies that improve the understanding of jurors. More research on the public communication of uncertainty could prove helpful on this point.[70]

The Collective Acceptance of Facts

The science court institutionalizes a critical public attitude toward scientific claims and gives a prominent role to organized dissent. Given the current climate of widespread distrust of expertise, some might think that we would be better served by an institution designed to increase deference to expert authority rather than one that encourages skepticism. But this objection blurs the distinction between questioning specific claims and rejecting expertise altogether.[71] Increasing appeals to the authority of experts or shielding scientific claims from public scrutiny is unlikely to solve the problem of expertise, and would further disempower the public and encourage irresponsible policy

---

70. Anne Marthe van der Bles, Sander van der Linden, Alexandra L. J. Freeman, and David J. Spiegelhalter, "The Effects of Communicating Uncertainty on Public Trust in Facts and Numbers," *Proceedings of the National Academy of Sciences* 117, no. 14 (2020): 7672–83.

71. For a discussion of the contrast between healthy skepticism and the conspiracist attack on knowledge-producing institutions, see Nancy L. Rosenblum and Russell Muirhead, *A Lot of People Are Saying: The New Conspiracism and the Assault on Democracy* (Princeton, NJ: Princeton University Press, 2019).

making. Far from promoting the rejection of expertise, the science court is built on the assumption that it is dangerous for democracies to deprive ordinary citizens of the opportunity to engage with the scientific expertise that forms the basis of many policies. An institution designed to give nonexperts the best possible chance of understanding and evaluating expert claims is an acknowledgment of the importance of the collective acceptance of facts, and a suggestion for how to achieve it without abandoning democratic values or expert knowledge. At a time where few facts can be taken for granted, this in itself is a major contribution of this proposal.

It is important to remember that the science court is not meant to solve all the problems that arise from the use of science in a democratic society on its own. It is meant to function as one part of a model designed to improve the use of science across the political system. I have argued that this model involves more openness and public dissent from scientific advisers, more responsible reporting from the media, and more public engagement with the uncertainty and incompleteness of scientific views. The next chapter turns to the question of how funding decisions for science could complement these, and considers the proper relationship between science and democracy at the funding stage.

5

Justifying Public Funding
for Science

"WAS DUCK PENIS study an appropriate use of taxpayer money?" ran a Fox News headline from 2013. The associated article attacked a government-funded animal behavior study on duck genitalia conducted at Yale University as a wasteful use of federal money. Two years earlier, another scientific study had been held up for ridicule: "Your tax dollars at work: shrimp on treadmills."[1] Although the immediate target of the attack was different, the main goal was the same: to criticize the National Science Foundation for wasting hundreds of thousands of taxpayer dollars in support of scientific projects that were supposed to seem obviously trivial to a layperson. To make the point more vivid, the earlier story included a rather fascinating video from the study, featuring—what else?—a shrimp exercising on a miniature treadmill, while a scientist took notes on its performance.

Mocking randomly selected examples of "silly" science has become a standard rhetorical tool for US Republicans who want to complain about the federal government's wastefulness.[2] In another line of attack, the Trump administration proposed major cuts to all federal programs supporting climate change research on the grounds that it clashed with the energy needs and

1. "Was Duck Penis Study Appropriate Use of Taxpayer Money?," Fox News, March 25, 2013, http://www.foxnews.com/opinion/2013/03/25/was-duck-penis-study-appropriate-use -taxpayer-money.html; "Your Tax Dollars at Work: Shrimp on Treadmills," Fox News, May 26, 2011, http://video.foxnews.com/v/960953334001/?#sp=show-clips.

2. Nell Greenfieldboyce, "'Shrimp on a Treadmill': The Politics of 'Silly' Studies," NPR, August 23, 2011, https://www.npr.org/2011/08/23/139852035/shrimp-on-a-treadmill-the -politics-of-silly-studies.

economic vitality of the country.[3] This time, scientists were charged with actively harming US economic interests. These attacks were part of a longer-term effort to delegitimize state support for scientific research and push for the privatization of science. Since the 1980s, there has been a steady increase in the proportion of research funded privately, especially in more applied and technology-driven areas.[4] Basic research is still mostly publicly funded, but it too is coming under pressure.[5]

The ability to fund new scientific research is a crucial power over the shape of knowledge in a society. I argued in chapters 2 and 3 that scientific knowledge sets the agenda for political decision-making, frames social conflicts, and determines the possibilities for political action. What the public and politicians consider to be desirable, feasible, or possible is shaped by the knowledge available to them. The absence of the right kind of knowledge makes it difficult to criticize existing policies and realize alternative visions of society. Funding decisions for science thus play a crucial role in determining the direction, speed, and limits of social and political change. By the time that scientific findings visibly affect citizens' lives, though—usually through their use in decision-making or new technologies—it is too late to change the contours of existing knowledge. The public and policy makers can accept or deny new scientific findings, but they cannot acquire a different body of knowledge or undo the science that has been done. To fully understand the relationship between science and democracy, we must therefore examine how decisions about science funding are made.

The current system of public funding for basic research was put in place following the close alliance between scientists and the state during World War II. This relationship was cemented with the provision of large amounts of funding for scientific research after the war. The unprecedented scale of public

3. Doyle Rice and Ledyard King, "Trump's Budget Proposal 'Savages' Climate Research, Scientists Say," *USA Today*, May 23, 2017, https://www.usatoday.com/story/news/nation/2017/05/23/trumps-budget-proposal-savages-climate-research-scientists-say/102062556/; Scott Waldman, "Trump Administration Seeks Big Budget Cuts for Climate Research, *Scientific American*, March 7, 2017, https://www.scientificamerican.com/article/trump-administration-seeks-big-budget-cuts-for-climate-research.

4. Daniel S. Greenberg, *Science for Sale: The Perils, Rewards, and Delusions of Campus Capitalism* (Chicago: University of Chicago Press, 2007); Philip Mirowski, *Science-Mart: Privatizing American Science* (Cambridge, MA: Harvard University Press, 2011).

5. On basic research as mostly publicly funded, see Shahar Avin, "Centralized Funding and Epistemic Exploration," *British Journal for the Philosophy of Science* 70, no. 3 (2017): 629–56.

investment in science intensified the mutual dependence between scientists and the state. It increased the power of science and technology to drive social, economic, and political transformations, and assigned scientists a leading role in the process. It also gave the state a significant stake in the conduct of scientific research and created the possibility of democratic control over the activities of scientists—a power that could be used with more or less restraint. This chapter focuses on the question of how a democratic society ought to wield this power, and strike a balance between the value of scientific autonomy and the right of citizens to have a say over the research they support through their taxes.

Science has a unique status among public goods in the degree of autonomy given to a professional community over the distribution of public funds. To explore the relationship between scientific autonomy and democratic control, we must understand how this unusual funding regime was justified in the first place, and ask whether existing justifications offer adequate resources for theorizing the relationship between the funding of science and the political influence of its products. This inquiry will allow us to rethink the proper level and kind of democratic input into funding decisions, and consider how else the division of labor between science and democracy may be drawn. Justification becomes necessary when the value of an activity can no longer be taken as self-evident. Research in the natural sciences continued to enjoy high levels of public support and financial investment from the state for many decades. This meant that while political theorists dedicated a lot of attention to defending funding for the arts, the humanities, and environmental preservation in the face of low or declining public support, the grounds for funding science received little philosophical scrutiny.[6] This chapter seeks to fill this gap.

I proceed by discussing two different justifications offered in support of funding science. I show why they are inadequate and then sketch a third line of justification that provides a fuller account of the political implications of science funding. I first focus on Bush's influential arguments from the 1940s; they laid the foundations for an ambitious program of publicly funded science in the United States. Bush maintained that universal public benefits would follow from basic scientific research, and that these benefits

6. Stefan Collini, *What Are Universities For?* (London: Penguin Books, 2012); Veronique Munoz-Dardé, "In the Face of Austerity: The Puzzle of Museums and Universities," *Journal of Political Philosophy* 21, no. 2 (2013): 221–42; Martha Nussbaum, *Not for Profit: Why Democracy Needs the Humanities* (Princeton, NJ: Public Square, 2016).

would be best realized if scientists were given a high degree of autonomy to pursue their curiosity. I then contrast Bush's arguments with Rawls's liberal justification. Rawls did not play a key part in shaping the national funding regime for science, but he provided one of the most influential normative accounts for how a liberal democratic state should treat the provision of public goods. His account implicitly rejected elitist approaches such as Bush's by arguing that state support for science (and art) must be based on the benefit principle, which would ensure that each person paid only for the benefits that they wanted.

Despite their differences, both these accounts failed to pay attention to the important political implications of scientific research: its role in determining political problems and their possible solutions, and its status as the authoritative source of knowledge for the modern state. I will argue that the close connection between scientific inquiry and truth, and special link between science and policy in the modern state, supply additional grounds for justifying the public funding of science beyond those that apply to the provision of other public goods, such as infrastructure or clean air. I will end by discussing which kinds of political input into funding decisions would be appropriate once we recognize the political consequences of funding science.

Bush and a Vision of the Common Good

At the end of World War II, President Franklin Delano Roosevelt asked Vannevar Bush to develop a new vision for how the government might support scientific research in the postwar period.[7] The report that Bush produced in response, called *Science: The Endless Frontier,* became the most influential document setting out the role that science could play in a large modern democracy.[8] Bush's argument had two key features. The first was the justification of public support for science almost entirely on the basis of the expected material benefits.

7. For more background on the debates around science in this period, see Daniel J. Kevles, *The Physicists: The History of a Scientific Community in Modern America* (1978; repr., New York: Knopf, 1995); Daniel L. Kleinman, *Politics on the Endless Frontier: Postwar Research Policy in the United States* (Durham, NC: Duke University Press, 1995); Alfred K. Mann, *For Better or for Worse: The Marriage of Science and Government* (New York: Columbia University Press, 2000); James D. Savage, *Funding Science in America: Congress, Universities, and the Politics of the Academic Pork Barrel* (Cambridge: Cambridge University Press, 1999).

8. Kleinman, *Politics on the Endless Frontier.*

Advances in science when put to practical use mean more jobs, higher wages, shorter hours, more abundant crops, more leisure for recreation, for study, for learning how to live without the deadening drudgery which has been the burden of the common man for ages past. Advances in science will also bring higher standards of living, will lead to the prevention or cure of diseases, will promote conservation of our limited national resources, and will assure means of defense against aggression.[9]

This was a clever strategy for addressing the dilemma facing science at the end of the war. The US public appreciated the role that scientists had played in winning the war, but lacked a nonmilitary vision that could justify continuing to spend large amounts of public funds on abstract scientific research. At the same time, cutting-edge science had become increasingly dependent on the continuation of large amounts of public funding, as military investment in science during the war had changed the nature and scope of scientific research. Bush's challenge was to come up with a persuasive narrative for what science could do to improve the lives of ordinary citizens in order to ensure continued public investment in basic research.

The second key tenet of the report was the necessity of granting scientists a high degree of autonomy from political processes and giving them control over the distribution of public funds.[10] Bush claimed that the public would benefit most from science if scientists were left free to pursue abstract research into areas that interested them: "Scientific progress on a broad front results from the free play of intellects, working on subjects of their own choice, in the manner dictated by their curiosity for exploration of the unknown."[11]

These two features, which defined the structure of science policy for the next several decades, gave science funding an unusual status among public goods. Although spending on science was justified on the basis of its expected public benefits, the public would have little say in how these benefits would come about. This special status faced a justificatory challenge: its plausibility depended on establishing the link between the public good and scientists pursuing their curiosity, but it was not obvious what could supply it.

---

9. Vannevar Bush, *Science: The Endless Frontier* (1945; repr., Washington, DC: National Science Foundation, 1960), 10.

10. For other discussions of this ideal, see Heather Douglas, *Science, Policy, and the Value-Free Ideal* (Pittsburgh: University of Pittsburgh Press, 2009); Philip Kitcher, *Science, Truth, and Democracy* (Oxford: Oxford University Press, 2001).

11. Bush, *Science*, 12.

Any account that claims that some means will be effective in furthering the public welfare must either presuppose knowledge of what constitutes it or include a procedure for how it should be determined. Bush's report developed a clear and concrete vision of what would constitute the public good for the United States in the postwar years: full employment, the production of goods and services to raise the standard of living, cheaper and better products to give the country an advantage in international trade, technological developments to increase agricultural productivity, and medical advances to cure disease. In short, a materialist conception of progress that emphasized the benefits from rapid economic growth and increased productivity. While this was certainly a vision of a *common* good, in the sense that it claimed the universal desirability of this bundle of material goods, it pushed forward an elite-driven vision instead of allowing a democratic one.

It is significant that Bush rested his argument on a particular conception of the public welfare rather than admitting the inherently political and contested nature of its determination. In doing so, he was trying to avoid two alternative lines of reasoning. The first was the democratic line that the public interest should be determined through the appropriate political procedures, and be responsive to the wishes and values of the public, whatever they might be. This would have paved the way for more political input into the direction of scientific research as well as the total level of funding, so Bush wanted to avoid it. The second alternative was to maintain that scientists should be given the authority to determine what would be in the public interest because of their superior wisdom and knowledge. This would have been an explicitly antidemocratic line, but it would also have been philosophically more coherent than Bush's position because it provided a straightforward justification for giving full autonomy to scientists. Instead, Bush put himself in the position of having to defend the bold and untested empirical claim that abstract scientific research would be the best way to maximize the public benefits that he had described.

Since no system of public funding for basic research of a comparable scale had ever been tried before, it was impossible to offer empirical evidence to support this claim. And Bush did not provide a direct explanation from scientific research to specific outcomes. Rather, his main point was that whatever people might want from science, basic research would be the way to get it.[12] Scientists would distribute funds among projects with an eye to solving

---

12. Kitcher, *Science, Truth, and Democracy*.

important puzzles, contributing to scientific progress, stimulating further research, or opening up new possibilities for future inquiry. Public benefits
would follow if scientists simply pursued the most significant advances and
produced the most important breakthroughs. The argument mimicked the
classical economists' argument for the free pursuit of private profit: just as the
economists claimed that society as a whole benefited from the pursuit of profit
by individual entrepreneurs, so too would society as a whole benefit from
individual scientists' free pursuit of their curiosity.

The high level of trust in science and scientists after the war meant that this
argument did not meet much resistance. But public acceptance does not
amount to normative merit; it is crucial to examine what could justify it. There
may not have been empirical evidence, but there were influential philosophical
arguments about scientific progress that had clear implications for the distribution of funds. Bush's claim that the greatest scientific progress would be
made if scientists were free to pursue their interests followed a theory of scientific progress developed by Michael Polanyi.

In a series of papers in the 1940s, Polanyi traced scientific progress to the
activities of a community of scientists sharing methods and standards, left free
from political interference.[13] He developed the argument through an analogy
between scientists and actors in a free market economy.

> [The] self-coordination of independent initiatives leads to a joint result
> which is unpremeditated by any of those who bring it about. Their coordi-
> nation is guided as by "an invisible hand" towards the joint discovery of a
> hidden system of things. Since its end-result is unknown, this kind of co-
> operation can only advance stepwise, and the total performance will be the
> best possible if each consecutive step is decided upon by the person most
> competent to do so.[14]

There were two key epistemic points supporting this claim. The first was
the indeterminacy of scientific research. Polanyi maintained that it was

13. Michael Polanyi, "Cultural Significance of Science," *Nature* 147 (1941): 119; Michael Po-
lanyi, "The Growth of Thought in Society," *Economica* 8 (1941): 428–56; Michael Polanyi, "The
Autonomy of Science," *Scientific Monthly* 60 (1945): 141–50; Michael Polanyi, "The Planning of
Science," *Political Quarterly* 16 (1945): 316–28. More developed versions of his views can be
found in later works such as Michael Polanyi, *The Logic of Liberty: Reflections and Rejoinders*
(Chicago: University of Chicago Press, 1951); Michael Polanyi, "The Republic of Science: Its
Political and Economic Theory," *Minerva* 1 (1962): 54–73.

14. Polanyi, "The Republic of Science," 55.

impossible to predict where the most significant scientific advances would come from. Any attempt by a funding committee to direct the course of science toward a specific purpose would fail because of this limitation. Note the striking similarity between this claim and Friedrich Hayek's argument against central planning on the basis of informational limitations.[15] Just as Hayek argued that the insurmountable information problem facing central planners showed the futility of government interventions into the economy, Polanyi contended that the indeterminacy of science meant that government interference with funding decisions would be pointless. His alternative was to leave it all to experts: "So long as each allocation follows the guidance of scientific opinion, by giving preference to the most promising scientists and subjects, the distribution of grants will automatically yield the maximum advantage for the advancement of science as a whole."[16]

The second point that held up the argument was a cumulative view of scientific progress. Polanyi subscribed to the traditional view of science moving incrementally toward a complete picture of the truth. He compared the scientific enterprise to a giant jigsaw puzzle, with each scientist carefully watching the moves of others in order to make the new moves that became possible as a result of earlier ones.[17] This account assumed a fundamental unity in science, such that all research fits together to form a coherent whole, which corresponds to the truth about the physical laws of the universe. The selection of research questions is thus not really an open choice; the scientist's problems are given by earlier work in the area and the gaps in existing knowledge. The significance of a research agenda comes from the role of the particular finding in filling out the missing pieces in the puzzle and contributing to its completion.[18] Science benefits society simply by making rapid progress in completing the puzzle.

The indeterminacy argument is hard to dispute, but does not support autonomy for scientists on its own. It simply shows the difficulty of directing the research enterprise toward predictable outcomes. Indeterminacy could just as

15. Friedrich August von Hayek, "The Use of Knowledge in Society," *American Economic Review* 35, no. 4 (1945): 519–30.

16. Polanyi, "The Republic of Science," 60.

17. Polanyi, "The Republic of Science."

18. For more on this philosophical position and its weaknesses, see John Dupré, "Science and Values and Values in Science," *Inquiry* 47, no. 5 (2004): 505–14; Kitcher, *Science, Truth, and Democracy.*

easily lend support to a distribution scheme that allocates funds equally among all projects or one that determines allocation randomly. Polanyi's argument for scientific autonomy depends on his second claim that scientific progress is a linear and cumulative movement toward a unified picture of truth. This view implies that there are only a few possibilities for new discovery at any given moment, and these are determined by the most recent discoveries in a particular area. The next move in the puzzle is fairly definite, but only scientists who have been closely following the work of other scientists can know what it is. Central planners who want to direct research toward the solution of pressing social problems will simply not be able to tell which areas are likely to yield discoveries. This justifies giving autonomy to scientists over decisions about the allocation of funds.

This picture of linear and cumulative scientific progress, however, was radically challenged by Kuhn in *The Structure of Scientific Revolutions.*[19] Kuhn affirmed Polanyi's description of everyday scientific research as an esoteric puzzle-solving activity, but rejected the claim that each problem solved was one small step toward a full picture of truth. He maintained instead that the ordinary activities of scientists, which he called normal science, should be understood as advancing a particular paradigm—a set of shared methods, standards, and accumulated knowledge. The puzzles selected by scientists are significant only relative to the paradigm rather than corresponding to an external standard of truth. The most significant and radical discoveries take place during periods where the paradigm becomes unable to solve important puzzles and breaks down. It is then replaced by a new one in a dramatic event that Kuhn called a scientific revolution. Normal science is still essential to significant discoveries, yet only because it prepares the conditions that make a revolution possible. The revolution itself is unintended and strongly resisted by practitioners of the old paradigm; they want to defend their own methods and findings.

An important consequence of Kuhn's challenge was to undermine the intrinsic significance of the everyday puzzles of scientists and thus also Polanyi's claim that scientists know best which problems are likely to produce major discoveries. Kuhn nonetheless agreed with Polanyi that scientists must be left to pursue their puzzles—not because they knew where the next major breakthrough would come from, but because only a scientist working within a paradigm could detect the anomalies that would eventually lead to a revolution. Paradoxically, the

19. Thomas Kuhn, *The Structure of Scientific Revolutions* (1962; repr., Chicago: University of Chicago Press, 2012).

necessary condition for radical and creative novelty turned out to be a scientific community that was rigidly controlled, esoteric, and elitist.

The problem with this position, as Paul Feyerabend pointed out, was that it took for granted the continued proliferation of alternative views capable of challenging a paradigm and precipitating a revolution, but did not explain why this should be true.[20] In fact, Kuhn's own view implies that the scientific community is structured precisely to extinguish this possibility: Kuhn claimed that scientists working in a paradigm would do everything in their power to resist threats to their paradigm. This led Feyerabend to argue that Kuhn was excusing the most dogmatic and narrow-minded features of postwar science.[21] It is therefore curious that Kuhn did not acknowledge the possibility that scientists might succeed only too well and end up extinguishing radical innovation. The risk that normal science might stifle innovation is even more serious under a system of scientific research that depends heavily on the availability of funding. If scientists working in paradigm have a high degree of autonomy over the distribution of funds, then the easiest way to defend a paradigm is by funding projects that develop the paradigm and not funding those that challenge it radically.

One of the most original contributions of Kuhn's *Structure* was to shift attention away from the lone individual following the scientific method, and to the dynamics of a community of scientists working with shared and unquestioned standards, norms, and assumptions. Yet Kuhn's famous examples of scientific revolutions were all drawn from periods that preceded the emergence of a highly professionalized community of scientists with shared and strongly enforced norms. Innovators such as Nicolaus Copernicus, Isaac Newton, Antoine-Laurent Lavoisier, and Albert Einstein, who overturned established scientific consensus, emerged in scientific contexts without a professional community with institutional tools for resisting new ideas. This mismatch between his examples and conclusions makes it difficult to share Kuhn's belief that a closed community of specialists entrusted with complete control over funding decisions will continue to produce radical ideas that undermine their own shared assumptions and findings. The argument for giving complete autonomy to scientists over funding decisions runs into difficulty

---

20. Paul Feyerabend, "Consolations for the Specialist," in *Criticism and the Growth of Knowledge*, ed. Imre Lakatos and Alan Musgrave (London: Cambridge University Press, 1970), 197–203.

21. Feyerabend, "Consolations for the Specialist."

when we consider the possibility that the process might develop in ways that would prevent rather than encourage the free pursuit of ideas and emergence of significant discoveries. If, as Kuhn's theory implies, a highly autonomous scientific community can use control over funding to resist new ideas that might threaten existing paradigms, outside intervention may be necessary to ensure the continued possibility of radical challenges.

Rawls's Liberal Justification

Bush made his case on the basis of a specific, rather materialistic conception of progress and flourishing for the postwar United States. But relying on such a particular vision of the good to support public funding for science in a democracy is controversial. After all, scientific research is not equally valuable to all people, and not all who value it do so for the instrumental reasons that Bush offered. Since dedicating large amounts of funds to science will detract from other goods and services that the state could be providing, we must consider whether and when it is acceptable for a state to fund an activity on the basis of a specific vision of the good. A promising way to address this question is by juxtaposing Bush's arguments for science funding with liberal ones that implicitly reject his approach. To this end, I will examine Rawls's arguments for funding science in a liberal society. The passages where Rawls mentioned science are few and invariably paired with public funding for the arts, but they are well worth studying as they provide a striking contrast to Bush's approach. While Bush appealed to a specific conception of the good, at the heart of Rawls's argument was the idea that justification for the public provision of certain kinds of goods should avoid appealing to any particular conception of what constitutes a public benefit.

In *A Theory of Justice*, Rawls divided expenditures by the state into two categories: those required by justice and those that are not. The former is governed by his two principles of justice and apply to the background institutions in a society, including legal definitions of property rights and a scheme of taxation.[22] These are expenditures necessary for the sustenance of a just basic structure in which all resulting distributions of income and wealth would also be just. These expenditures are not subject to a popular vote. What justifies their imposition on all citizens is the fact that they are a necessary cost of living in a

22. John Rawls, *A Theory of Justice* (Cambridge, MA: Belknap Press of Harvard University Press, 1971), 62, 29.

just, mutually advantageous cooperative venture.[23] What is of interest here is Rawls's second category of public goods: those not required by justice.

Rawls pointed out that the requirements of justice might not cover all public expenditures that citizens might wish to make. "If a sufficiently large number of them find the marginal benefits of public goods greater than that of goods available through the market, it is appropriate that ways should be found for government to provide them."[24] Since justice does not require the provision of these additional public goods, the principle regulating their provision should be solely that of benefit; individuals should be taxed in proportion to the benefits they receive. Interestingly, the only specific class of public goods that Rawls mentioned as an illustration for this category was funding for the arts and sciences.[25]

Rawls gave more precise form to this requirement by appealing to Knut Wicksell's unanimity principle. In an 1896 article, Wicksell had argued that if a public good is an efficient use of social resources, there must be a distribution of tax burdens that would gain unanimous approval. Decision makers should consider proposals for public goods together with alternative schemes for the distribution of tax burdens. Only those tax schemes that gained unanimous approval should be provided. Under such a scheme, those who would derive no benefit from the good would not be forced to pay, and the distribution of burdens across individuals would track the value of the good for each person.[26] Rawls adopted this principle and proposed the creation of a separate branch of government—the exchange branch—to deal with its application to particular decisions.[27]

Wicksell's unanimity principle is strange as a theory of public goods.[28] For one thing, it ignores the possibility of strategic behavior or bargaining by individuals to secure better deals for themselves. Under this system, individuals

---

23. This contentious claim gave rise to the famous debate between Rawls and Robert Nozick on whether the benefits derived from a cooperative enterprise could ground an obligation to share its burdens. Robert Nozick, *Anarchy, State, and Utopia* (New York: Harper and Row, 1974).

24. Rawls, *A Theory of Justice*, 282.

25. Rawls, *A Theory of Justice*, 331.

26. Knut Wicksell, "A New Principle of Just Taxation," in *Classics in the Theory of Public Finance*, ed. Richard Musgrave and Alan Peacock (London: Palgrave Macmillan, 1958), 72–118.

27. Rawls, *A Theory of Justice*, 283.

28. David Miller, "Justice, Democracy and Public Goods," in *Justice and Democracy: Essays for Brian Barry*, ed. Keith Dowding, Robert E. Goodin, and Carole Pateman (Cambridge: Cambridge University Press, 2004), 131–33; Richard Tuck, *Free Riding* (Cambridge, MA: Harvard University Press, 2008), 191–92.

have an incentive to misrepresent their preferences in order to secure a lower tax rate for goods that they would like to have provided. Since everyone has this incentive, the unanimity principle will result in the under provision of public goods. This is a version of the classic free rider problem. But even if we bracket the possibility of strategic behavior, the rule still allows for an extremely narrow scope for the state provision of nonjustice goods. Specifically, it only allows the provision of goods that constitute a Pareto improvement. Under this rule, the state could not make anyone subsidize goods that they would not benefit from or pay more for a good than its value to them. Only those taxation packages under which no one would be a net loser would pass the unanimity requirement. The principle applies a narrow understanding of economic efficiency to the realm of public provision.

Rawls justified his adoption of this principle on the grounds that it would prevent the state from imposing unwanted burdens on people by appealing to perfectionist justifications that they did not share. While expenditures required by justice are justified on the basis that everyone benefits from a just system, this logic cannot be applied to discretionary goods, which are justified by appeal to particular conceptions of the good. "The principles of justice do not permit subsidizing universities and institutes, or opera and the theater on the grounds that these institutions are intrinsically valuable and those who engage in them are to be supported even at some expense to others who do not receive compensating benefits."[29] This, he argued, would be equivalent to forcing people to subsidize the private expenses of others.[30]

In *Justice as Impartiality*, Brian Barry rejected the view that a market-mimicking procedure must be the solution for disagreement over public goods and maintained instead that the decision should be settled through a democratic

29. Rawls, *A Theory of Justice*, 325, 332.

30. This became a fundamental tenet of liberal conceptions of state funding. In the 1990s, when US conservatives were attacking government funding for the National Endowment of the Arts, there was a lively debate among liberal theorists on whether public support for the arts could be justified on a liberal conception of the state, assuming that support for "high" art went against the preferences of the majority. Most theorists concluded that it was difficult to do so. On this intriguing debate, see Harry Brighouse, "Neutrality, Publicity, and State Funding of the Arts," *Philosophy and Public Affairs* 24, no. 1 (1995): 35–63; Noël Carroll, "Can Government Funding of the Arts Be Justified Theoretically?," *Journal of Aesthetic Education* 21, no. 1 (Spring 1987): 21–35; Ronald Dworkin, *A Matter of Principle* (Cambridge, MA: Harvard University Press, 1985), 225; Joel Feinberg, "Not with My Tax Money: The Problem of Justifying Government Subsidies for the Arts," *Public Affairs Quarterly* 8, no. 2 (1994): 101–23.

process.[31] On Barry's procedural account, decisions about public goods must be treated the same way as any other political decision, where people with different and incompatible preferences must reach an agreement about what to do. Appeals to specific conceptions of the good would be allowed as arguments in deliberation, and the final decision would be made through a fair decision procedure, such as majority rule, agreed on in advance.

Replacing unanimity with a majoritarian decision rule means that some people would be forced to subsidize goods that they did not want and would not benefit from. What justifies imposing tax burdens on the minority in these cases is not the intrinsic value of the good but instead the fact that its provision is agreed on through a fair decision procedure, which gives no special advantage to any conception of the good. This, in turn, is justified on the basis of the overall desirability of a system that allows individuals to cross-subsidize public goods for others: I subsidize your football stadium in return for you subsidizing my opera house. As long as each person has a reasonable chance of finding themselves in the majority some portion of the time, everyone has reason to prefer this system since it will supply more of the goods that each person desires.[32]

Rawls shifted to a similar position in *Political Liberalism* and *Justice as Fairness: A Restatement*.[33] While he maintained the earlier distinction between goods that concern "constitutional essentials and questions of basic justice" and those that do not, he argued in these later works that the provision of nonjustice goods could be decided by a democratic vote rather than through unanimity. With the requirements of justice already in place, citizens could try to persuade each other of their preferences over public goods with arguments drawn from their comprehensive doctrines. Rawls was reluctant to put many goods in the nonjustice category, but art and science remained paradigm cases: "Fundamental justice must be achieved first. After that, a democratic

31. Brian Barry, *Justice as Impartiality: A Treatise on Social Justice, Vol 2* (Oxford: Oxford University Press, 1995), 143–51.

32. This is obviously an imperfect system. There might well be a problem with persistent minorities that never get any of their desired goods. Miller points out that in an earlier work, Barry suggested that a majoritarian decision rule would be chosen only in societies where people could expect to find themselves in the majority at least half the time. Brian Barry, "Is Democracy Special?," in *Democracy and Power: Essays in Political Theory, Part 1* (Oxford: Clarendon Press, 1991), 24–60; Miller, "Justice, Democracy and Public Goods."

33. John Rawls, *Political Liberalism* (New York: Columbia University Press, 1996), 214; John Rawls, *Justice as Fairness: A Restatement* (Cambridge, MA: Harvard University Press, 2001), 151–52.

electorate may devote large resources to grand projects in art and science if it so chooses."[34]

Rawls did not elaborate on the particular arguments that individuals could make in favor of funding science. As long as they did not derive from claims about justice, their content would be irrelevant to the structure of justification. Bush's appeal to the necessity of basic scientific research for economic development, full employment, and progress would be equally acceptable as appeals to the intrinsic value of knowledge and understanding. His justification did not discriminate between the content of particular conceptions of the good or particular accounts of benefit. This framework could accommodate a vision such as Bush's if most individuals were convinced of the desirability of such a vision of public welfare and the role of abstract science in achieving it. But this view must compete on the political terrain with rival ones, such as those that defend a more targeted and applied science, or those that claim that the state need not support science at all.

This line of reasoning will naturally have trouble providing the normative grounds for continued provision in cases where most individuals do not want the benefits and do not take the activity to be intrinsically valuable. This difficulty has plagued liberals trying to defend state support for the arts against conservatives attacking it on the grounds that it wastes money, and encourages highbrow or offensive art.[35] How might the liberal argue against the charge that certain scientific studies are wasteful? One possible response is to point out the overall advantages of a system of basic research that involves funding many esoteric and seemingly trivial projects. Since we cannot know in advance which projects will yield the most innovative discoveries, it is reasonable to diversify funding. Even if most projects turn out to be dead ends, the system can be justified by the significant advances made in some areas. Instead of scrutinizing each project on the basis of whether it contributes to the desired public benefits, this approach contends that we should judge the system.

The second defense shows how particular studies are in fact indirect means to achieving the desired public benefits. A requirement that grant applications outline expected public benefits encourages scientists to justify their work in terms of imagined downstream benefits even where these may not be obvious.

---

34. Rawls, *Justice as Fairness*, 152.

35. See Brighouse, "Neutrality, Publicity, and State Funding of the Arts"; Carroll, "Can Government Funding of the Arts Be Justified Theoretically?"; Dworkin, *A Matter of Principle*; Feinberg, "Not with My Tax Money."

The scientists who conducted the duck genitalia and shrimp treadmill studies followed these two strategies to defend their work against attacks. The author of the duck study emphasized the importance of funding basic research, while the author of the shrimp study underscored the links between the health of marine organisms and safety of the seafood that humans consume.[36] While the first defense appeals to the value of the overall practice of funding basic research, the second is a direct defense of specific projects.[37]

Both of these arguments will have limited reach, though. The appeal to the benefits of the overall system may shield individual projects from demands for justification, but the system as a whole must still be justified, and opponents may well reject its value. Moreover, appealing to the system cannot always remove individual projects from scrutiny either. If an area of research is particularly expensive and removed from the benefits that the public wants, citizens or their representatives can rightly object that this goes beyond what they believe is justified by appeal to the overall system of funding basic research, and special justification must be provided because of the heavy tax burden. In the end, appeals to the benefits may prove unpersuasive to critics, and the government may decide to withdraw funding from a specific project, as in the highly publicized case of the Superconducting Super Collider project in the 1990s.[38]

I have called this a limitation, but it counts as such only from the perspective of someone trying to defend science funding against attacks. This limitation need not be a bad thing from a democratic perspective; it might simply point to the right place to draw the line between scientific autonomy and political interference in the undertaking of costly scientific projects with taxpayer money. In any case, my goal here is not to settle the question of which scientific projects are worth their cost but instead to sketch the form that normative arguments for and against funding might take within a framework of private

36. Patricia Brennan, "Why I Study Duck Genitalia," *Slate*, April 2, 2013, http://www.slate .com/articles/health_and_science/science/2013/04/duck_penis_controversy_nsf_is_right _to_fund_basic_research_that_conservatives.html; David Scholnick, "How a $47 Shrimp Treadmill Became a $3-Million Political Plaything," *Chronicle of Higher Education*," November 13, 2014, https://www.chronicle.com/blogs/conversation/2014/11/13/how-a-47-shrimp -treadmill-became-a-3-million-political-plaything.

37. For a discussion of the distinction between justifying a practice and justifying a particular action falling under it, and its significance for utilitarianism, see John Rawls, "Two Concepts of Rules," *Philosophical Review* 64, no. 1 (January 1995): 3–32.

38. Kevles, *The Physicists*.

conceptions of the good. Ultimately, an institution justified on the basis of its public benefits must be supported by evidence of the benefits or their likelihood, whether the benefit is interpreted on a case-to-case basis or at a systemic level. If we endorse the liberal view that people should not be forced to pay for benefits they do not want, then the question of whether public funds should be spent on science will depend on the ability of defenders to persuade opponents of the value of funding it.

A Democratic Argument for Funding Science

The Rawlsian argument puts science squarely in the category of zoos and lacrosse fields: goods that are justified by privately held conceptions of the good. There might be good reasons to value them, but large costs should not be forced on people who do not want the benefit. But does science properly belong in this category? Attacks against science on the basis of triviality and wasting money appear to confirm its status as a privately valued good. Yet science has been the target of another line of attack mentioned earlier: that certain areas of scientific research are harmful to national interests. Recent attacks on climate change funding, for instance, have taken this form. This charge stems from the claim of science and scientists to be providing truths about the world. Scientific findings hold the power to change the beliefs that are reasonable to hold and public policies that are reasonable to pursue. This creates significant political stakes around the outcomes of research.

Appeals to private benefits or private value cannot account for these political implications. The problem is not that the private benefits framework gives the "wrong" answer; it might well turn out that there are in fact no persuasive normative arguments for continued public funding when a majority does not wish to spend its money on research that it regards as harmful or biased. The problem is rather that conceptualizing funding for science as a matter of benefits for individuals misses the irreversible collective impact of scientific findings. The close connection between scientific inquiry and truth, and special link between science and policy in the modern state, create additional reasons for the public funding of science that go beyond those that apply to the public provision of roads, bridges, or clean air. Recognizing the political consequences of scientific research is crucial not only for a more robust defense of continued public funding but also for an appreciation of what role, if any, political interventions might properly play in decisions about the allocation of the public funds set aside for science.

Before I develop this argument, two clarifications are in order. First, I take it to be a sociological fact that science occupies the role of an authoritative source of truth in secular modern states. Even those who reject specific scientific claims do so on the grounds that they are wrong or unscientific, instead of denying the claim of science to providing authoritative and useful knowledge. I recognize that this need not be the case. The political status of science does not simply follow from its claim to revealing the truth (or the truth of this claim) but rather from the fact that this is widely accepted. Astrology also claims to reveal the truth, but has no policy influence, and it is conceivable that scientists might be allowed to pursue their research in a theocratic society without being regarded as a reliable source of truth for policy purposes.

Second, I accept the view that the state shouldn't impose costs on people for public goods that are justified by appeal to privately held conceptions of the good that are neither shared nor determined through democratic procedures. This rules out a justification of public funding for science that focuses on its essential role for a certain vision of human flourishing and excellence—one that might appeal to the importance of the pursuit of truth for the proper development of human capacities, or the possibility of leading deeper and more complex lives. Instead, I will make the case from a set of fundamental political interests that are shared by all who participate in a democratic society. The argument is consistent with a liberal framework, but provides a fuller account of how the role of funding for science could be theorized within it. In particular, I challenge Rawls's categorization of science funding as a discretionary good whose justification appeals purely to private benefits or private value.[39]

With that, I want to turn to the question of how the connection between science, truth, and democracy affects arguments for public funding. I will focus on three distinctly political roles for science: as a means for effective policies, resource for the empowerment of citizens, and agenda setter for political debates.

---

39. One way to situate my argument within a Rawlsian framework would be to say that science should fall within the domain of public reason—of fundamental questions about constitutional essentials and basic justice—rather than that of discretionary goods. The difficulty with this is that the political role of science is not about justice per se but rather about the requirements of a well-functioning democratic process. Perhaps Rawls's bipartite classification scheme of justice and nonjustice goods is too restrictive for this purpose. Introducing a third category of goods that can be justified by appeal to political values more broadly might be a useful amendment.

I have already showed how modern states depend on scientific expertise. Scientific knowledge enables policy makers to find effective means for realizing democratically determined ends. In areas ranging from public health to environmental protection, from technological risk to foreign policy, science is instrumental in the making of good policies. It is a core assumption of modern states that claims to truth in policy contexts must be scientific. Since politicians and government officials cannot produce the knowledge that they need, they depend on the existence of a scientific community capable of producing and sharing it. This establishes an intimate relationship between the activities of scientists at research institutions and the fundamental political interest in bringing about good outcomes and attaining collectively determined ends.[40] Dewey recognized this relationship when he noted that "genuinely public policy cannot be generated unless it be informed by knowledge, and this knowledge does not exist except when there is systematic, thorough, and well-equipped search and record."[41]

This reason for funding science is instrumental; it stems from a shared public interest in good outcomes. Most political issues cannot be addressed only by appealing to values. On questions such as whether human activities contribute to climate change, smoking causes cancer, or a virus is transmitted through aerosols, citizens and their representatives depend on scientists to provide the facts—and the facts matter crucially. This role that science plays in politics is not specifically democratic; an authoritarian regime could likewise justify large amounts of public investment in scientific research on the grounds that it would improve outcomes.

Democracies, unlike nondemocracies, have a further political interest in funding scientific research because of the part that scientific inquiry can play in providing citizens with a source of knowledge independent from the state that they can use to hold government officials accountable. At the very least, democratic accountability requires a sphere of free public discourse. On complex technical issues, however, commonsense knowledge is unlikely to be enough

40. This applies even more strongly to the social sciences, although I bracket them here. Desmond King shows how the social science funding regime in Britain was built on the assumption of the identity of publicly funded research with political ends. Desmond King, "Creating a Funding Regime for Social Research in Britain: The Heyworth Committee on Social Studies and the Founding of the Social Science Research Council," *Minerva* 35, no. 1 (1997): 1–26. By contrast, the US National Science Foundation tended to favor social science research that looked more like basic science. See Otto N. Larsen, *Milestones and Millstones: Social Science at the National Science Foundation, 1945–1991* (New Brunswick, NJ: Transaction Publishers, 1992).

41. John Dewey, *The Public and Its Problems* (Chicago: Swallow Press, 1927), 178.

to give citizens a meaningful ability to check the activities of government officials since decisions are often justified by appeals to expert knowledge. Ordinary citizens must also have access to expertise to be able to understand and challenge policy makers. Scientific inquiry can fulfill this need, and support a specifically democratic form of competence and empowerment for citizens. Publicly available scientific knowledge can allow citizens to revise their opinions, form more informed preferences, and reconsider their ends. These are related to good outcomes, but they possess greater normative weight because of their relationship to the essential condition of democratic legitimacy that citizens have a meaningful opportunity to hold policy makers accountable.

The two arguments provided so far focus on the role of scientific knowledge in the making and judging of political decisions. Yet scientific inquiry also plays an important role in determining which issues will arrive at the decision stage in the first place. As I argued in chapter 2, scientists' decisions about which questions to pursue determines which issues will acquire political salience, and their theoretical and methodological choices in answering a question shape how the public and politicians will think about it. On issues such as COVID-19, climate change, GMOs, vaccines, and environmental degradation, scientific findings have set the political agenda and defined the terms of public debate. For instance, early studies of climate change prioritized identifying causal mechanisms and making predictions, with the assumption that solutions would follow more readily from an understanding of the underlying physical processes.[42] Funding for research on the possibilities for adaptation and technological innovation lagged behind funding for research into the study of the physical science for many years. This limited the options considered feasible and desirable for tackling the problem, even while placing the fact of climate change and its impacts irreversibly on the political agenda. Similarly, the systematic absence of scientific knowledge in an area can keep significant issues off the political agenda altogether. Scholars have documented how the lack of research on the harms of industrial processes in chemicals, energy, food, and transportation technologies made the mobilization of public opposition more difficult, and strengthened the hand of corporate interests trying to prevent the politicization of these issues.[43]

42. Daniel Sarewitz, "Normal Science and Limits on Knowledge: What We Seek to Know, What We Choose Not to Know, What We Don't Bother Knowing," *Social Research* 77, no. 3 (2010): 997–1010.

43. David J. Hess, *Undone Science: Social Movements, Mobilized Publics, and Industrial Transitions* (Cambridge, MA: MIT Press, 2016).

Those who make the funding decisions can rule out certain courses of action and avenues for change, intentionally or unintentionally. This makes funding decisions for science the locus of an important political power. Science can enhance democratic rule if the knowledge that becomes available supports democratic priorities and increases a society's ability to shape its future through favored courses of action. But it can also constrain democratic possibilities if new knowledge thwarts collectively determined goals and aspirations, or creates unforeseen and potentially unwanted needs and problems. Whether the relationship between science and democracy will be a productive one, then, depends not only on the quality and reliability of the knowledge but also on the decisions about which knowledge is pursued and how.

Let me consider what these three arguments for the political role of scientific research imply for funding decisions and contrast this with Bush's recommendation for giving a high degree of autonomy to scientists to attain the greatest material benefits for the public.

First, the political impact of scientific knowledge and especially its instrumental role in bringing about certain outcomes may place some areas of science funding in the category of goods whose public provision is required by justice. For example, if there is a duty of justice to help those who are harmed by the natural disasters caused by climate change, funding climate research may be necessary for the ability of the state to fulfill this duty. Similarly, if the state has a duty to provide health care for those with certain rare diseases, this might require funding research into the discovery of cures. The argument follows something like a transitivity principle for duty: if you have a duty to do x, you also have the duty to do those things that are a means to x.[44] In practice, it will often be highly unclear that a particular area of research will yield the right kind of information to fulfill a duty of justice. It will be uncertain whether funded projects will succeed or do so in time, whether the findings will turn out to help the cause of justice rather than create new and unimagined injustices, and so on. But in cases where certain avenues of research can plausibly be tied to research necessary to realize duties of justice, there will also be a duty to publicly support these areas of inquiry.

Second, the political role of science has implications for how decisions about the distribution of funds should be made. If the democratic benefits of

<hr/>

44. As Robert Goodin puts it, "You ought to do things that are means towards the principal thing for the same reason you ought to do that principal thing." Robert Goodin, "Excused by the Unwillingness of Others?," *Analysis* 72, no. 1 (2012): 18–24.

science are at least part of the justification for the public support of science, then institutions that make funding decisions should be designed with an eye to realizing the desired benefits. Which decision structures will best realize these goals is ultimately an empirical question, but in the absence of the right sort of data and the difficulty of testing alternative institutional structures, we need to rely on theoretical principles to guide the process of institutional design. The agenda-setting power of science and its effect on the possibilities for democratic rule suggests that priority-setting decisions should involve some democratic input. How funds should be distributed between, say, biomedical research and environmental studies, or space exploration and oceanography, is a decision that must be made by appealing to the values and preferences of citizens. These are analogous to fundamentally political questions about how to distribute funds between education and health care, or national defense and environmental quality. Since science is supported by public funds for the purpose of public benefit, however that is to be defined, the priorities of the scientific research agenda should be set with democratic input. Expert opinion on the likelihood of making significant progress in these scientific areas will be relevant to the decision, of course, but in the end, the ordering of priorities must be made democratically. Indeed, the current practice in the United States is to shape priorities for science funding on the basis of national political priorities.

The significant increase in the share of scientific research funded by corporations, philanthropists, and private foundations since the 1980s raises the worry that the interests, needs, and priorities of corporations and private individuals are shaping the scientific agenda.[45] This means that private organizations can circumvent political processes by enacting their vision of a good society through their private funding decisions as opposed to seeking majority support. The increase in privately funded science therefore has implications for the distribution of public funding. The distributive impact of scientific projects becomes particularly salient when more science is privately funded. If scientific issues that benefit certain groups or industries are supported disproportionately through private funds, then it will be necessary to counterbalance the effects of private science through more directed public funding. The distribution of public funds for science is a powerful tool that

---

45. William J. Broad, "Billionaires with Big Ideas Are Privatizing Science," *New York Times*, March 15, 2014, http://www.nytimes.com/2014/03/16/science/billionaires-with-big-ideas-are -privatizing-american-science.html?_r=0; Greenberg, *Science for Sale*; Mirowski, *Science-Mart*.

remains in the hands of citizens in their collective capacity, which they can use to influence the direction of science-driven social change in the face of increasing privatization.

It is common to draw the line for political input into science funding at the general level of priority setting, leaving the distribution of funds within each area to scientists. This is inadequate because it leaves the determination of how a particular issue will be considered in the public sphere and the alternatives that will be available entirely to scientists. Formal decision-making power may still lie in political processes, but it is constrained by funding decisions made earlier by scientists. The distribution of funds within an area has different political implications than priority setting at a more general level. While priority setting determines which problems will gain more traction, the distribution of funds within an area determines the range of possible answers to a particular problem. At the decision stage, laypeople can either accept one of the available scientific options or reject them all, but they cannot produce new science. The success of democratic deliberation about how to act depends on the availability of competing views that citizens and their representatives can examine and challenge. Citizens will not have a meaningful opportunity for choice unless they are presented with a wide range of alternatives on the same question.

The agenda-setting role of science points to the need for more democratic input into funding decisions. The next question is what the other two arguments—from the interest in better outcomes to the interest in empowering citizens to hold their government accountable—imply about the desirability of political intervention into funding. One plausible answer is that both supply prima facie reasons for insulating funding decisions from outside interference. The success of the relationship between scientific inquiry and competent policy making depends on the scientific community's ability to set internal standards of quality. The distribution of funding among competing research proposals is one of the main ways in which the scientific community discriminates between good ideas and bad ones. The overall success of this gatekeeping mechanism in providing reliable knowledge justifies the reliance of policy makers on the findings of scientists. As long as it is accepted that scientists are more likely to possess the methods that will lead to truth, democracies have an interest in protecting scientific funding from outside attempts to steer it. Political interference with funding decisions can impose myopic preferences over long-term commitments. Worse, it may be motivated by a desire to prevent good policies by blocking the emergence of truth. This also effectively prevents citizens from acquiring the information that they need to criticize government policies. When a government decides to withdraw funds from research on a pressing political issue such as climate change, it is reasonable to suspect that the interference is motivated by a desire to suppress politically inconvenient truths and thereby prevent sensible policy making on the issue.

These arguments support a prima facie case for scientific autonomy over the distribution of public funds. The case is only prima facie because the use of funding as a gatekeeping mechanism for quality also provides a reason *against* scientific autonomy over the distribution of funds. Recall the earlier discussion of Kuhn's claim that scientists left to pursue their own puzzles without interference would prepare the necessary conditions for the most radical discoveries, even as they actively resisted new ideas that challenge the assumptions of their paradigm. I pointed out that under a system where most scientific research depends on large amounts of funding, those who control funding decisions might succeed only too well in rejecting radical new ideas that would lead to the most significant discoveries. If scientists working within a paradigm have control over the distribution of funds, then the easiest way to ensure the continued success of the paradigm is by funding projects that extend the paradigm and not funding those that challenge it.[46]

The role that funding has come to play in scientific research means that the possibility of dissent in science depends not only on the absence of constraints on free inquiry but also on an active funding strategy that supports dissenting views and distributes funds among a wide variety of approaches. To leave the decision entirely to scientists' assessment of what counts as good quality in light of existing standards can prevent the funding of new ideas that can potentially challenge those very standards and expose the errors of widely accepted scientific views. Polanyi compared scientists to actors in a market economy whose uncoordinated pursuit of truth would bring about the best results for all. To extend his analogy, giving scientific agencies full control over the distribution of funds could lead to the emergence of monopolies in knowledge production, led by the assumptions and priorities of those who sit on funding committees. It is therefore necessary to establish funding institutions that can fulfill the function of antitrust law for science and ensure fair competition among scientific ideas.

---

46. For evidence that this is happening, see Joshua M. Nicholson and John P. A. Ioannidis, "Conform and Be Funded," *Nature* 492 (2012): 34–36.

To make these arguments more concrete, let me discuss their institutional implications. I proposed a two-tiered role for democratic influence over science funding: a direct role in setting research priorities to ensure that scientific growth responds to democratic aims, and an indirect stake in promoting diversity and dissent to encourage novelty and significant discoveries. The implication of the first is straightforward and largely consistent with current practice: representatives must develop a vision of how science can advance the public good and select the issue areas that must be prioritized for funding, taking distributive and justice-based concerns into account. They should be responsive to public opinion on research priorities, and maintain clear and accessible channels for public input and accountability.

The main challenges on this point are practical. Citizens and their representatives must be able to assess how well scientific research realizes collectively determined aims, and which funding efforts deliver significant and socially valuable results. They cannot make sound judgments on the appropriate level and direction of funding without evidence and feedback on the results of past spending. While the Rawlsian view would simply aggregate the benefits received by individuals, a theory that rests on a collective conception of benefits requires a public process for defining, measuring, and assessing collective benefits. The problem is that there are currently no systematic efforts to measure the effectiveness of publicly funded research in advancing democratically determined conceptions of the public interest—or any other measures of success, for that matter.[47]

I propose a large-scale effort to survey the evidence on scientific outputs, focusing on the success of research in delivering expected or unexpected public benefits. This would be complicated and subjective, to be sure, but having imperfect studies would still be preferable to taking benefits entirely on faith. Note that this is not a demand for evidence of short-term and purely material benefits, contra those making wastefulness charges and driving the push for privatization. It is a call for a more holistic study of the value created by publicly funded science over the decades, granting that it can take a long time for scientific results to materialize, and the progress of science involves failures

---

47. In *Science-Mart*, Mirowski attempts the task of assessing whether the quality and quantity of scientific outputs suffered due to the commercialization of science in recent decades, but does not include a metric for public value. Economist Mariana Mazzucato's *The Entrepreneurial State* is one example of what such an effort could look like, though it does not aim to provide a full survey. Mariana Mazzucato, *The Entrepreneurial State* (London: Demos, 2011).

and dead ends. More systematic evidence of the value of public science could also galvanize public support and advocacy for scientific research at a time when it is coming under attack.

The implication of my second argument—about the democratic stake in ensuring diversity and preventing the monopolization of funding—is less obvious. I argued that scientists' interest in defending their own paradigms creates the risk that they will resist and possibly extinguish more innovative approaches. I also suggested, contra Polanyi, that the indeterminacy of science does not support giving scientists complete autonomy over funding. Neither of these claims, however, implies that anyone else would be more effective than scientists in diversifying funding and promoting radical ideas. Nonexperts may have incentives to support novelty and diversity, but they are unlikely to be able to distinguish reliably between highly technical proposals. Moreover, majoritarian decision-making procedures are not usually the best method for supporting minority views.

What we need are institutional arrangements designed to support different approaches, particularly unconventional projects, but without presuming the ability to predict and plan the future of research. Recent proposals for using lotteries in the distribution of science funding fulfill both these criteria. Scholars have defended the use of lotteries on the grounds that peer review is not sufficiently reliable in assessing the quality of proposals, favors low-risk projects, displays gender and racial bias, and involves high time costs.[48] My argument here provides an additional reason for their desirability: lotteries would eliminate concerns about scientists favoring their own paradigms at the risk of suppressing criticism and novelty. The difference between researchers working under dominant paradigms and those who challenge it would be irrelevant in a lottery; each would have an equal chance of funding. This equality would constitute a significant improvement in the chances of unusual or high-risk proposals and make it preferable to use lotteries to distribute some portion of funds, at least among proposals that meet a certain quality threshold in an initial expert screening. Evidence that peer review is not sufficiently successful in ranking the quality of proposals would provide additional reasons in favor.[49]

---

48. Avin, "Centralized Funding and Epistemic Exploration"; Shahar Avin, "Mavericks and Lotteries," *Studies in History and Philosophy of Science Part A* 76 (2019): 13–23; Ferric C. Fang and Arturo Casadevall, "Research Funding: The Case for a Modified Lottery," *mBio* 7, no. 3 (2016): 1–7; Daniel S. Greenberg, "Chance and Grants," *Lancet* 351 (1998): 686.

49. Nicholas Graves, Adrian G. Barnett, and Philip Clarke, "Funding Grant Proposals for Scientific Research: Retrospective Analysis of Scores by Members of Grant Review Panel,"

Other institutional innovations to support the same goals might include earmarking funds for unusual approaches and less established scientists, ensuring a diversity of scientific viewpoints on funding committees, reserving seats for nonexperts and experts from different scientific disciplines, and supporting long-term research in more speculative projects. It should not be surprising that all these changes would be good for science as well as democracy. The point is that the democratic justification for publicly funding research places some of the responsibility for setting up and overseeing such institutions on policy makers and the public rather than giving full autonomy to the scientific community over the use of public funds. This is not an argument for handing over particular funding decisions to bureaucrats, politicians, or ordinary citizens but instead for democratic responsibility and oversight for the creation and maintenance of a system of funding that encourages competition, diversity, and dissent within the scientific community.

A Shared Responsibility

In order to determine the proper role for democratic input into the level and distribution of public funds for science, we first need to understand how the current science funding regime has been justified and then ask whether these justifications are satisfactory. Different arguments for supporting science with public funds suggest different answers for the acceptability of cuts to funding or political interventions with how it is distributed. To answer these questions, I first examined Bush's arguments for setting up a vast federally funded scientific enterprise after the war on the grounds that it would boost economic progress and productivity. I then turned to Rawls's more modest liberal argument for funding science as a voluntary good, which could be provided by the state if individuals found the benefits desirable on their private conceptions of the good. The problem with both of these views, I argued, is that they neglected the political consequences of funding science. Neither can help us understand what is truly at stake in something like the view that if climate change research appears to harm US economic interests, there should simply be less funding for it.

To account for the political stakes in funding science, I sketched an alternative justification, rooted in the shared democratic interests of citizens: bringing about good policy outcomes, setting the political agenda, and acquiring the

BMJ 343 (2011), https://www.bmj.com/content/343/bmj.d4797; Fang and Casadevall, "Research Funding."

knowledge and competence to hold policy makers accountable on technical issues. These three interests pull in different directions on the desirability of political input into funding decisions. The agenda-setting power of science points to the necessity of more democratic input, both in setting priorities and diversifying approaches within issue areas, especially in a society with increasingly privatized science. The need for reliable knowledge for policy making and accountability points in the direction of giving autonomy to scientists, but must be balanced against the worry that a closed expert community might develop patterns of funding that stifle dissent and innovation, which in turn would limit the opportunities for contestation in the public sphere.

The question of how to set up funding institutions that encourage dissent and facilitate the emergence of truth is usually thought to be a concern for scientists and philosophers of science. One of my goals here was to show that this problem is a democratic one too, and the responsibility for resolving it must be shared between policy makers and scientists. This is not only because scientific research is pursued with public funds and invariably justified by appeal to public benefits but also because the direction of new scientific research, reliability of scientific findings, and possibility of dissent within the scientific community have direct impact on democratic deliberation and decision-making. In the end, I hope to have made it clear that public funding for science is a deeply political issue, from the justification of the decision to support science with public funds to the making of specific decisions about how to distribute funds—and rightly so.

6

Dangerous Science and the
Limits of Free Inquiry

IN SUMMER 2015, over a thousand artificial intelligence and robotics experts signed an open letter calling for a preemptive ban on the development of autonomous lethal weapons—so-called killer robots. The letter argued that the deployment of these weapons would become feasible in a matter of years and their use would pose a serious threat to humanity.[1] "We do not have long to act," declared another letter from the founders of robotics companies. "Once this Pandora's box is opened, it will be hard to close."[2] The ensuing debate among researchers and tech experts had two key features. The first was the high uncertainty of possible outcomes. The likelihood of harms and benefits was unknown, and the outcomes envisioned were often speculative. The second was a dispute about the wisdom of banning research at an early stage rather than addressing the harmful effects of the technology as they became clearer. Critics who opposed the ban argued that it would be better to regulate the technology once it became available for use as opposed to foreclosing possible benefits by interfering with its development.[3]

---

1. "Autonomous Weapons: An Open Letter from AI and Robotics Researchers," https://futureoflife.org/open-letter-autonomous-weapons. As of this writing, the letter has 30,717 signatories.

2. "An Open Letter to the United Nations Convention on Certain Conventional Weapons," https://futureoflife.org/autonomous-weapons-open-letter-2017.

3. Kenneth Anderson and Matthew C. Waxman, "Debating Autonomous Weapon Systems, Their Ethics, and Their Regulation under International Law," in *The Oxford Handbook of Law, Regulation and Technology*, ed. Roger Brownsword, Eloise Scotford, and Karen Yeung (Oxford: Oxford University Press, 2016), 1097–118; Michael Schmitt and Jeffrey Thurnher, "Out of the Loop: Autonomous Weapon Systems and the Law of Armed Conflict," *Harvard National*

These features characterize the most interesting and controversial cases of high-risk, high-uncertainty new technologies, ranging from killer robots to geoengineering and heritable gene editing. These technologies involve complex scientific research, and their possible effects are insufficiently understood. They offer significant benefits but also pose serious risk of harm, with plausible mechanisms whereby their use could lead to catastrophe. Proponents of these technologies emphasize their revolutionary potential, while detractors focus on the dangers.[4] On each of these issues, concerns about the potentially catastrophic harms from the use or misuse of the technologies have triggered calls for halting research.[5] Although moratoriums or outright bans on research remain rare, concerned scientists and NGOs have made appeals to restrict scientific research on several issues over the past two decades. Besides killer robots, the targeted issues have included geoengineering, heritable gene editing, synthetic biology, nanotechnology, human cloning, gene drives, lethal flu viruses, and part-human, part-animal chimeras. The increasing frequency of these appeals makes the justifiability and desirability of restricting high-risk scientific research a pressing issue.

The dangers evoked by these research areas illustrate the sheer magnitude of the effects that science and its technological products can have on society. They also reveal how these effects are shaped by earlier decisions about the selection of research questions, which are made by scientists who may neither intend nor condone the eventual societal risks. The question of how to respond to

Security Journal 4, no. 2 (2013): 231–81; Ronald Arkin, "Lethal Autonomous Systems and the Plight of the Noncombatant," *AISB Quarterly* 137 (2013); Thomas Simpson and Vincent Müller, "Just War and Robots' Killings," *Philosophical Quarterly* 66, no. 263 (2016): 302–22. For a general argument for giving up attempts to control technology, see Kevin Kelly, *What Technology Wants* (New York: Penguin, 2010).

4. Clare Heyward, "Is There Anything New under the Sun? Exceptionalism and Novelty in Debating Geoengineering Governance," in *The Ethics of Climate Change Governance*, ed. Aaron Maltais and Catriona McKinnon (London: Rowman and Littlefield, 2015), 135–54.

5. Eric S. Lander, Françoise Baylis, Feng Zhang, Emmanuelle Charpentier, Paul Berg, Catherine Bourgain, Bärbel Friedrich, et al., "Adopt a Moratorium on Heritable Genome Editing," *Nature* 567, no. 7747 (2019): 165–68; Mike Hulme, *Can Science Fix Climate Change? A Case against Climate Engineering* (Cambridge, UK: Polity Press, 2014); ETC Group, "UK Royal Society on Geoengineering: The Emperor's New Climate?," press release, August 28, 2009. Alan Robock has called for a ban on geoengineering experiments outdoors, while Edward Parson and David Keith proposed a ban on experiments above a certain threshold. Alan Robock, "Is Geoengineering Research Ethical?," *Security and Peace* 30, no. 4 (2012): 226–29; Edward Parson and David Keith, "End the Deadlock on Governance of Geoengineering Research," *Science* 339, no. 6125 (2013): 1278–79.

dangerous research not only raises ethical dilemmas but poses a challenge for the relationship between science and democracy too. While it is uncontroversial to suggest that new technologies involving serious risk of harm should be regulated or even banned, it remains unusual to argue that earlier stages of scientific inquiry may be targeted as the appropriate site of restrictions. By the time the results of research materialize in the shape of a risky technology, though, it is often too late to prevent the damage. This problem forces democratic societies to strike a difficult compromise between their commitment to two deeply held principles: freedom of inquiry and the need to protect citizens from harm.

This chapter asks whether it is permissible for a democratic society to restrict or ban scientific research, and if so, under what conditions. My goal is not to develop a complete ethical theory that specifies which scientific projects may or must be restricted but instead to offer a framework for deciding whether and when to restrict research under conditions of empirical and normative uncertainty. I examine the strongest reasons against the permissibility of restrictions and suggest some principles to guide the decision process, concentrating in particular on the appropriate interaction between expert-led and democratic processes. The argument starts from the ethical framework for the regulation of scientific research with human subjects and proceeds by offering modifications to adapt it to the purpose of governing new technologies. Two main questions arise in the process: the first is whether it is justifiable to impose restrictions at the research stage to prevent harms that will arise during the application of a technology, and the second is whether it is justifiable to restrict research preemptively on the grounds of public fear and anxiety, before there is sufficient evidence establishing the risk of harm. I answer both questions in the affirmative and then defend this position against objections.

Defining the Scope

The argument in this chapter centers on new technologies that hold the promise—or threat—of a dramatic transformation. These technologies typically involve complex scientific knowledge, and their possible effects are insufficiently understood. Despite this, or perhaps because of it, they are believed to provide a solution to important social problems even as they pose a threat of serious harm and possibly even catastrophe. Many of these effects are imagined, and uncertainty about the impacts is a defining feature of these cases. These technologies frequently capture the popular imagination, and inspire books and films with postapocalyptic narratives.

To be more precise, I define this category of new technologies as exhibiting the following features:

1. The development and use of the technology, or its application, poses a serious risk of harm while promising significant benefits, and there are scientifically plausible mechanisms whereby it could lead to catastrophe
2. There is uncertainty about possible outcomes; not only is there insufficient information to assign probabilities to outcomes, but there is significant uncertainty about what the set of possible outcomes is
3. The use of the technology is expected to have broad and potentially irreversible societal impacts, which will not be limited to those who choose to use the technology

Although these criteria are open to interpretation, they are intended to capture an intuitive difference between technologies involving local and more predictable risks, such as toxic substances or new household appliances, and technologies whose risks are of a greater magnitude and whose impacts are expected to be more dramatic.[6] The distinction is probably a matter of degree rather than of kind given that many new technologies involve a risk of harm, uncertainty about outcomes, and broad impacts. But a significant difference in degree matters; it can make more radical measures appropriate as well as justify different institutional arrangements and decision-making procedures for dealing with the consequences. In the following, I will be concerned with cases that lie on the more dramatic end of the spectrum. Certain geoengineering technologies and killer robots are paradigmatic cases in this category.

Geoengineering: This refers to the deliberate large-scale manipulation of the earth's natural systems to counteract the effects of climate change. The term is used to describe a variety of different technologies; I am interested here in one of the more promising yet more radical technologies called stratospheric aerosol injection (SAI), which involves the injection of sulfate aerosols into the

6. This bears some similarity to what Nick Bostrom has called existential risks, but I put less emphasis on the existential character of the threat. See Nick Bostrom, "Existential Risks: Analyzing Human Extinction Scenarios and Related Hazards," *Journal of Evolution and Technology* 9, no. 1 (2002).

stratosphere to reflect a fraction of incoming sunlight back into space, thereby creating a sunshade for the ground beneath.[7] The major promise of SAI is that it could slow down the worst impacts of climate change, including global temperature increase and the melting of sea ice and glaciers. At the same time, there are major risks associated with it. Sulfate particles could damage the ozone layer, change precipitation levels and monsoon cycles, cause droughts and cyclones, disrupt established ecological systems, and alter agricultural production patterns.[8] Furthermore, any halt in deployment could lead to sudden and rapid warming, which means that the decision to deploy might entail a commitment to continual interventions. Changes in temperature and precipitation are also likely to vary across regions, thus potentially exacerbating existing regional inequalities. Once the technology is developed, the possibility of deployment by private or state actors to harm opponents cannot be ruled out. The use of weather modification to attack political opponents has historical precedent. During the Vietnam War, for instance, the United States used rain inducement techniques to swamp North Vietnamese supply lines and suppress antiwar protests.[9] Because SAI involves modifying the planetary climate, it would not be possible to avoid its effects.

Killer robots: Lethal autonomous weapons, popularly known as killer robots, combine artificial intelligence with machinery capable of deadly force.[10] While the automated weapons currently used by militaries can only execute predefined tasks set by humans, killer robots have the ability to perceive and process features of their environment, and use this information to solve a problem or reach a goal. On the battlefield, this usually involves the ability to move

7. Royal Society, *Geoengineering the Climate: Science Governance and Uncertainty*, policy document 10/09, 2009.

8. For detailed discussions of the risks of SAI, see Wil Burns and Andrew Strauss, eds., *Climate Change Geoengineering: Philosophical Perspectives, Legal Issues, and Governance Frameworks* (Cambridge: Cambridge University Press, 2013); Hulme, *Can Science Fix Climate Change?*; Alan Robock, "20 Reasons Why Geoengineering May Be a Bad Idea," *Bulletin of the Atomic Scientists* 64, no. 2 (2008): 14–18.

9. Robock, "20 Reasons Why Geoengineering May Be a Bad Idea"; James Fleming, *The Checkered History of Weather and Climate Control* (New York: Columbia University Press, 2012).

10. Robert Sparrow, "Killer Robots," *Journal of Applied Philosophy* 24, no. 1 (2007): 62–77; Isaac Taylor, "Who Is Responsible for Killer Robots? Autonomous Weapons, Group Agency, and the Military-Industrial Complex," *Journal of Applied Philosophy* (2020), doi.org/10.1111/japp.12469.

around, identify targets, open fire, and kill. While automated weapons can be fully controlled by a human, who is ultimately responsible for its behavior, the ability of killer robots to learn new behavior and form new beliefs in a context allows them to act unpredictably, thereby moving them beyond human control. Those who defend killer robots argue that their use might reduce casualties and collateral damage in war.[11] Nevertheless, the possibility that killer robots might unpredictably and unaccountably kill large numbers of people cannot be ruled out. The open letter cited earlier argued that these weapons would be ideally suited for use in assassinations, subduing populations, and ethnic violence. The authors also pointed out that the development of killer robots by a single country would precipitate a global arms race, with the result that these weapons would eventually fall into the hands of terrorists, dictators, warlords, and rogue nations, with potentially catastrophic results for humanity.[12]

In the following, I will examine the normative issues raised by the possibility of restricting these kinds of high-risk, high-uncertainty technologies. I use the phrase "restricting research" to refer to the decision to pause or halt research in an area through the imposition of a moratorium or indefinite ban. Regulations intended to greatly slow down the pace of research activity, such as publication restrictions that prevent other researchers from recreating experiments, may also be considered in this category. These measures can be enforced by governments or voluntarily observed by scientists. The choice of policy tools would be determined by feasibility and efficacy considerations.

Harms to Subjects versus Harms from Application

Political interventions to limit research may sound problematic, but there are already significant constraints on what researchers may or may not do to advance scientific knowledge. The most serious constraint is on the grounds of potential harm to human subjects. The currently accepted practice is that scientific experiments that pose serious harm to human subjects are not

---

11. Anderson and Waxman, "Debating Autonomous Weapon Systems"; Schmitt and Thurnher, "Out of the Loop"; Arkin, "Lethal Autonomous Systems"; Simpson and Müller, "Just War and Robots' Killings."

12. "Autonomous Weapons."

allowed.[13] All research proposing to use human subjects and conducted at research institutions receiving federal funding must be approved in advance by an institutional review board (IRB) that scrutinizes proposals for compliance with safety and welfare regulations.[14] This is no trivial restriction of the kinds of knowledge that scientists are allowed to seek. Many areas of potentially beneficial medical research are ruled out because they are impossible to conduct without testing dangerous experimental treatments on human subjects.

These rules reflect carefully considered views on how to weigh the potential risks and benefits of scientific research, focusing especially on the trade-off between the scientific and social value of the knowledge to be gained from research with the harm inflicted in the process. They set out the conditions under which the value of the pursuit of knowledge cannot be enough to justify the risks inflicted in the process. They also have the advantage of being widely accepted. Although the Belmont report, which sets out ethical principles for the protection of human subjects, was accepted as late as in 1979, the need for such regulation has come to appear self-evident today. For both these reasons, they provide a helpful starting point for thinking about the regulation of risky new technologies.[15]

The Belmont report outlines three principles to protect human subjects in research: respect, beneficence, and justice. The respect principle requires experimenters to treat participants as autonomous agents. It is typically expressed in a requirement for researchers to obtain informed consent and ensure that conditions for the voluntariness of consent are met. The beneficence principle requires researchers to protect the well-being of the subject. It asks them not to harm subjects, and make sure that the research design maximizes

13. National Commission for the Protection of Human Subjects of Biomedical and Behavioral Research, "The Belmont Report: Ethical Principles and Guidelines for the Protection of Human Subjects of Research," US Department of Health and Human Services, April 18, 1979.

14. For discussions of the problems with IRBs, see Philip Hamburger, "IRB Licensing," in *Who's Afraid of Academic Freedom?*, ed. Akeel Bilgrami and Jonathan Cole (New York: Columbia University Press, 2015), 153–90; Judith Jarvis Thomson, Catherine Elgin, David A. Hyman, Philip E. Rubin, and Jonathan Knight, "Research on Human Subjects: Academic Freedom and the Institutional Review Board," *Academe* 92, no. 5 (2006): 95–100.

15. For another piece that takes this framework as a starting point, see David R. Morrow, Robert E. Kopp, and Michael Oppenheimer, "Toward Ethical Norms and Institutions for Climate Engineering Research," *Environmental Research Letters* 4, no. 4 (2009): 1–8.

the potential benefits and minimizes the potential harms. The justice requirement demands that the distribution of harms and benefits be fair. In practice, it is typically interpreted as applying to subject selection. It aims to ensure that vulnerable populations are not exploited in risky research, and wealthy populations do not receive favorable treatment in beneficial interventions.

One limitation of these rules is that they focus almost exclusively on harms inflicted during the research process. IRBs do not take into consideration the potential harms to individuals or society from the application of the findings, nor are they allowed to consider the risks that potential subjects would face if they did not participate in the experiment.[16] For instance, the life expectancy of a patient under available treatments cannot be considered when determining permissible levels of risk. The review process is thus insulated from the background conditions in which the research will be conducted. Broader impacts are considered only in order to ensure that the social value of the knowledge generated is sufficiently high to justify exposing subjects to a risk of harm. In the case of new technologies such as SAI and killer robots, however, the most serious risks do not arise during the research process itself but instead from the use of the resulting technology in a particular context. Considering the technology in the abstract or with reference only to the intentions of its developers will yield a partial view of the potential consequences. A proper evaluation of the effects of the technology must include both the risks from the responsible and controlled use of the technology, and the harms from its possible misuse in the wrong hands.

It may be clear that the governance of new technologies cannot ignore the foreseeable harms from use and misuse. Yet it is unusual to suggest that scientific research may be constrained to prevent harms that will arise from use. There is a strong intuition that since knowledge is never intrinsically harmful, regulation should target the application of findings, not the research. As one scientist puts it, "Science tells us how the world is. That we are not at the centre of the universe is neither good nor bad, nor is the possibility that genes can influence our intelligence or our behavior. Dangers and ethical issues only arise when science is applied in technology."[17] Current IRB rules support this view, as they rest on a clear distinction between harms inflicted on human

16. Ezekiel Emanuel, David Wendler, and Christine Grady, "What Makes Clinical Research Ethical?," *JAMA* 283, no. 20 (2000): 2701–11.

17. Lewis Wolpert, "Is Science Dangerous?," in *Scientific Freedom*, ed. Simona Giordano, John Cotton, and Marco Cappato (London: Bloomsbury, 2012), 31–42.

subjects and harms from the use of knowledge. We must therefore pause to ask what justifies this distinction and whether it is morally relevant.

There are two plausible justifications for maintaining the distinction. The first is the assumption that new knowledge gained from research will always be beneficial for society. This may well be true of the kinds of research that IRB rules were designed to regulate. It might be safe to assume, for instance, that knowledge gained from clinical research always improves our understanding of human health and well-being. Research aimed at the development of a new technology, however, is different because the mission of producing and using the technology drives the research. Expanding our understanding of nature is not the primary goal. The benefits and harms from the knowledge are often linked to the benefits and harms from the use of technology.

This leaves the second explanation: that harms from the use of technology in a particular context are in some way different from harms inflicted on human subjects. If the magnitude and probability of a risk of harm is held constant, it is difficult to see why the permissibility of harm would depend on whether scientists inflict it during the research process or others inflict it once the knowledge is used in a context. The intuition that there is a meaningful difference between the two is most likely driven by the immediacy of bodily interventions in clinical trials. SAI field experiments provide a counterexample that can help shake this intuition. There isn't a clear dividing line between large-scale experiments conducted by scientists to find out more about the effects of the technology and deployment of the same technology by the government to counter the effects of climate change.[18] The important variables that must be considered in deciding the permissibility of injecting sulfate aerosols into the stratosphere are the magnitude of the harms, their level of certainty, and their distribution.

There is, though, one significant difference between harms to subjects and harms from use for the purpose of determining whether a particular line of research should be allowed: in the former, the prohibition targets those who would have inflicted harm, while in the latter, researchers are targeted for the harms that might come about from the way other people use their research. Their freedom is limited to prevent others' wrongdoing. While an individual's freedom may be limited to prevent them from harming others, targeting a person for harms that

18. Edward Parson and Lia Ernst, "International Governance of Climate Engineering," *Theoretical Inquiries in Law* 14, no. 1 (2013): 307–38; Alan Robock, Martin Bunzl, Ben Kravitz, and Georgiy L. Stenchikov, "A Test for Geoengineering?," *Science* 327, no. 5965 (2010): 530–31.

they neither inflict nor intend might seem unfair. The acceptability of limiting the freedom of researchers depends on whether they can be held morally responsible for the downstream consequences of their research.

Some theories maintain that people cannot be held morally responsible for the unintended but foreseeable harm that results from others' responses to their actions. These views separate the moral rightness of an action from its consequences and do not adjust moral obligations in anticipation of the predictable actions of others. In doing so, they confer a special status to the person who ultimately inflicts the harm, such that they and only they can be held responsible for the result, regardless of what has taken place before.[19] What could possibly justify treating acts at earlier stages of a causal chain as irrelevant from a moral perspective? The one coherent answer requires accepting the Kantian view that each moral agent is an uncaused cause, and no other person or action can be taken to have caused their voluntary actions. If this were true, then researchers could not be held morally responsible for the use of their research, and restricting their freedom would be unjustified. But this is a strange metaphysical view, and it is difficult to find good reasons why researchers should be relieved of moral responsibility when they have contributed to a foreseeable bad result.

The crux of the matter lies instead in the scope of reasonable foresight. Scientists cannot be expected to foresee all uses of their research, and many effects of new technologies are truly unpredictable. The widely acknowledged value of free inquiry also weighs against making the burden of foresight onerous. Yet the technologies discussed here are defined as posing serious risk of harm, with scientifically plausible pathways to catastrophic outcomes. The risks and catastrophic possibilities are, by stipulation, foreseen. Moreover, the use of a technology is an entirely foreseeable outcome of research toward its development. The fact that researchers may not intend its use and even end up arguing against it is not enough to release them from responsibility for creating the risks.[20] In such cases of serious and foreseeable risks, the moral responsibilities of scientists can be understood through the categories of recklessness or negligence, which are defined by foresight or foreseeability.[21]

19. Michael J. Zimmerman, "Intervening Agents and Moral Responsibility," *Philosophical Quarterly* 35, no. 141 (1985): 347–58.

20. In fact, scientists are often the first to speak up against the use of technologies that they have helped develop, most likely because they feel responsible.

21. Heather Douglas, "The Moral Responsibilities of Scientists (Tensions between Autonomy and Responsibility)," *American Philosophical Quarterly* 40, no. 1 (January 2003): 59–68;

Recklessness involves the knowing imposition of foreseen and unjustified risks on others, while negligence involves a failure to foresee risks that were reasonably foreseeable. That researchers should be assigned some moral responsibility does not mean that they should be held equally responsible as those who make the decision to use the technology. Nor do those who use the technology become any less responsible simply because researchers bear some responsibility too. Responsibility is not zero sum in this way.

This argument may still not persuade those who insist that individuals should not be held morally responsible for the unintended consequences of their actions. An alternative and less controversial strategy might be to emphasize the distinction between accounts of moral and legal responsibility. The effort to draw the boundaries of permissible regulatory interference with research cannot be insensitive to consequences since regulation is concerned with the effects of new technologies on the welfare and protection of individuals in a social context. Since preventing harm is one of the main purposes of regulation, restricting certain kinds of actions can be justified on the grounds that doing so would be the most effective way to prevent foreseeable harms to others.

Of course, what the state may legitimately prevent an individual from doing is not entirely independent from questions about moral responsibility. For the state to be justified in interference, it needs to be the case both that there be a threat of harm to others and that the individual whose actions are prevented be in some way responsible for the harm. But responsibility can be interpreted more broadly in a legal and political context, and need not map onto responsibility according to any particular theory of individual morality. While assigning strict liability without fault is relatively rare, the legal categories of recklessness and negligence can be more capacious than their moral counterparts since the law has latitude in setting standards of care and reasonable foresight.

I have maintained that research into the development of new technologies may be prohibited not only on the basis of harms caused during the research process but also on the basis of a risk of harm from the use of the technology. This would require considerations about whether the harm in question is substantial enough, the risk of harm outweighs the possible benefits, and the restriction of individual liberty is unduly great. Under conditions of uncertainty, it would be reasonable to demand clear and substantial evidence for short-term harms, while accepting plausible scenarios of long-term harms that might not

Jonathan Wolff, "Risk, Fear, Blame, Shame and the Regulation of Public Safety," *Economics and Philosophy* 22 (2006): 409–27.

be as well established. The point is that the claims that knowledge itself can never harm and technology is intrinsically neutral are not enough to rule against the permissibility of restricting research.

A Democratic Equivalent of Consent

As noted earlier, the Belmont report stipulates a respect principle, which asks researchers to seek the informed consent of participants in order to ensure that subjects are treated as autonomous agents. I will now take up the question of what role, if any, something like the respect principle might play in the regulation of new technologies. The applicability of the respect principle to the regulation of new technologies requires justification because of the difference in scope between a lab experiment with easily identifiable participants and a technology such as SAI or killer robots whose effect on specific individuals remains extremely uncertain. Two questions in particular must be answered. The first is whether there is a need for consent at all. The current practice is for expert committees to decide whether and when to impose moratoriums on research. The second question is whether subjective determinants of risk, such as collective fears around a new technology, can be acceptable grounds for a democratic refusal to consent. In order to answer these questions, it is important to first clarify what the informed consent requirement is meant to achieve and how it interacts with the beneficence principle.

The beneficence principle aims to protect subjects from potential harms that could be inflicted during the research process. It prohibits researchers from inflicting harms above a certain magnitude, and then typically allows a cost-benefit analysis to ensure that the risk of harm during the research is proportionate to or outweighs the potential benefits to subjects.[22] Permitted projects must not only be safe but also must be oriented toward maximizing the well-being of the participants. This principle contains enough ethical content that it could regulate the entire process. An expert review board could complete all the necessary ethical calculations and deliberations to decide whether to approve a project. Why, then, is it necessary to require the autonomous choice of the participants?

There are at least two possible answers, which reflect different reasons for valuing autonomy.[23] The first is that it is valuable for individuals to make their

---

22. Emanuel, Wendler, and Grady, "What Makes Clinical Research Ethical?"

23. I follow the conception of autonomy developed in Joel Feinberg, *The Moral Limits of the Criminal Law Volume 3: Harm to Self* (New York: Oxford University Press, 1989); Joseph Raz, *The Morality of Freedom* (Oxford: Clarendon Press, 1986).

own choices, regardless of whether the choices are good or bad. Even if participating in an experimental treatment is objectively good for a person, it is valuable for them to have the ability to choose not to do so for whatever reason. This respects choice simpliciter. The second is that individuals often know better what is good for them and should be given the freedom to determine the standards by which goodness should be judged. Regardless of what an IRB judges about the safety or benefit of a project, individuals may not want to participate if it does not fit with their values, or it causes great fear, anxiety, or discomfort. Subjects may be more risk averse than the review board or more worried about certain kinds of risks. What constitutes the subjects' best interest or an acceptable level of harm are determined at least in part by their values, beliefs, fears, and aversions.

The beneficence and respect principles together secure a sphere of free choice, which is circumscribed by an expert assessment of what constitutes a permissible research proposal. This ideal regulatory structure cannot be carried over easily to the regulation of new technologies because respecting each individual's right to consent to a technology would mean disallowing the development or use of the technology altogether. Once the research for a technology is available, individuals cannot reject its impacts. At the same time, giving each person a veto over the development of new technologies would be an unreasonable constraint on the freedom of research and development. When individuals cannot opt out without constraining the freedom of others, the situation must be conceptualized as one that requires a collective decision about how to strike a balance between the freedom of researchers and the welfare and safety of others.

The regulation of technologies such as SAI and killer robots is more similar in this respect to the regulation of industrial pollutants than it is to the regulation of research with human subjects. While subjects in a biomedical trial can refuse to participate without preventing the experiment from going forward, each citizen is not granted the right to give or refuse consent to a pollution level because doing so would give too much power to individuals over the determination of what others may or may not do. Note that this difference is not due to the magnitude of the harm involved; toxic wastes could cause greater harm to the health of residents near an industrial plant than a scientific experiment might to potential participants.

The question, then, is what role, if any, remains for a procedure that captures the spirit of consent in the case of collective harms. One answer is that consent simply drops out, and experts direct the process on the basis of potential harms. This is the standard practice for the regulation of everyday risks such as industrial pollutants and toxic substances; it is also the approach used to regulate some high-risk scientific research. For instance, research aimed at developing more lethal forms of viruses was put under a moratorium by a National Institutes of Health expert panel in 2014 on the grounds that the risk of an accidental pandemic was too great to justify any benefits.[24] In a similar case, prominent biologists involved in the development of the gene editing technique known as CRISPR have called for a moratorium on its use in heritable genes.[25] The paradigm case for this approach remains the 1975 Asilomar Conference, where biologists decided to put a moratorium on research involving recombinant DNA.[26]

The consequence of dropping consent is that the process becomes entirely expert driven. A procedure that requires both a harm and consent principle ensures that both experts and nonexperts have power over the decision. Experts determine if a proposal is safe, or safe enough, and nonexperts decide if they want to accept or reject it. This structure gives priority to the harm principle, since no research deemed harmful by an IRB can go forward, but it still grants a crucial veto power to nonexperts. Without the consent principle, however, the process relies entirely on expert calculation and judgment. It derives its normative force from the reliability of the calculations about risks and benefits, and denies other citizens a chance to participate in the decision.

Why might this be a problem? Recall the two reasons given above for why it is important to require consent over and above expert determinations of harm. The first is that choice may be intrinsically valuable, even if individuals end up making decisions that are not in their best interest. The second is that individuals may judge the risks and benefits differently than an expert committee, such as because they are more or less risk averse, or fear certain kinds of risks more than others. Cultural theorists of risk have shown that risk attitudes are closely tied to collective moral systems and the organization of social relationships. Risks acquire salience according to social ascriptions of blame, trust,

24. Donald G. McNeil Jr., "A Federal Ban on Making Lethal Viruses Is Lifted," *New York Times*, December 19, 2017. https://www.nytimes.com/2017/12/19/health/lethal-viruses-nih.html.

25. Lander et al., "Adopt a Moratorium on Heritable Genome Editing."

26. Paul Berg, David Baltimore, Sydney Brenner, Richard O. Roblin, and Maxine F. Singer, "Summary Statement of the Asilomar Conference on Recombinant DNA Molecules," *Proceedings of the National Academy of Sciences of the United States of America* 72, no. 6 (June 1975): 1981–84.

responsibility, vulnerability, and fairness.[27] Behavioral studies have also demonstrated that individual perceptions of risks are not reducible to the magnitude and probability of the risk, and qualitative aspects of risks play a significant role in determining subjective responses. For instance, most individuals fear dramatic, uncontrollable, irreversible, or involuntary risks far more than everyday, controllable, reversible, and voluntary ones of the same expected value.[28]

Cultural and psychological factors can interact too.[29] In the context of high-risk new technologies, fears about loss of control and irreversible effects can be exacerbated by the ascription of recklessness, hubris, or even malice to the pursuit of these technologies, thus jointly amplifying the social impact of the bad outcomes foreseen by scientists. As a result, the social responses to technologies can end up being dominated by the fear of things going wrong.[30] Irreversible changes in the environment that destroy local ecological systems, rapid warming and natural disasters from failed geoengineering experiments, artificial intelligence that goes rogue and destroys humans, and killer robots that fall into the hands of terrorist groups—such scenarios of technological disasters can dominate the public imagination. The unpredictability of outcomes may create resistance to new technologies that go beyond what is supported by the scientific evidence, and can only be explained by people's feelings of fear, anxiety, and potential loss.[31]

27. Mary Douglas and Aaron Wildavsky, *Risk and Culture: An Essay on the Selection of Technological and Environmental Dangers* (Berkeley: University of California Press, 1982); Steve Rayner, "Cultural Theory and Risk Analysis," in *Social Theories of Risk*, ed. Sheldon Krimsky and Dominic Golding (Westport, CT: Praeger, 1992), 83–115; Wolff, "Risk, Fear, Blame, Shame and the Regulation of Public Safety."

28. Paul Slovic, Baruch Fischhoff, and Sarah Lichtenstein, "Facts and Fears: Understanding Perceived Risk," in *Societal Risk Assessment: How Safe Is Safe Enough?*, ed. Richard C. Schwing and Walter A. Albers (New York: Plenum Press), 181–214.

29. Roger E. Kasperson, Ortwin Renn, Paul Slovic, Halina S. Brown, Jacque Emel, Robert Goble, Jeanne X. Kasperson, and Samuel Ratick, "The Social Amplification of Risk: A Conceptual Framework," *Risk Analysis* 8, no. 2 (1988): 178–87; Nick Pidgeon, Roger E. Kasperson, and Paul Slovic, eds., *The Social Amplification of Risk* (Cambridge: Cambridge University Press, 2003); Ortwin Renn, *Risk Governance: Coping with Uncertainty in a Complex World* (London: Earthscan, 2008).

30. Royal Society, *Geoengineering the Climate*.

31. Gareth Davies, "The Psychological Costs of Geoengineering: Why It Might Be Hard to Accept Even If It Works," in *Climate Change Geoengineering: Philosophical Perspectives, Legal Issues, and Governance Frameworks*, ed. Wil Burns and Andrew Strauss (Cambridge: Cambridge University Press, 2013), 59–80.

We take it to be unproblematic that an individual who severely fears a medical procedure involved in a clinical trial should have the right to refuse consent. The collective equivalent of refusing consent to an IRB-approved project would be a democratic decision to ban a certain line of research, even where experts have deemed the evidence for harm to be insufficient. Would such a constraint on research be permissible? The answer depends on whether we think public policy ought to be responsive to subjective determinants of risk, such as collective fears and anxieties, especially in cases where expert risk assessment is inconclusive about the likelihood of harm.

For individual agents, the ability to act on fear is crucial for autonomy, regardless of the rationality of the fear. Autonomous agents control their own life and pursue projects of their own choosing; being in a state of fear and anxiety poses an obstacle to doing so. If a deep-seated fear is not removed, the agent may become powerless and alienated. The ability to act on fear is crucial for the ability to make decisions and pursue goals. The ideal of respecting autonomy would be incoherent if it were limited only to the circumstances in which there was sufficient evidence that the agent's feelings and decisions were rational. Acting on emotions rather than facts is a natural part of human agency, and so is making mistakes. These arguments also apply to collective decision-making. Democracy holds the promise that people can live in light of collectively determined values, and shape the world according to their needs, desires, and aversions. Fears and anxieties, along with hopes and desires, are crucial signs of what citizens take to be important to them; fears can reveal what they value the most.[32] Acting to remove a potential source of harm can be an instantiation of the democratic power to shape society in light of collective values as well as a precondition for the democratic ability to realize a positive vision of society thereafter. The development of some new technologies might bring about such significant social, political, and environmental changes that the decision about whether research should go forward may be better conceived as a decision about whether citizens agree to live in this altered world. This is a fundamentally political rather than scientific question.

To be clear, my argument is not that prohibiting a new technology on the grounds that it causes fear is always the right thing to do or that it should be the first recourse. The reasonable initial step is to evaluate whether the fears are

32. Martha Nussbaum, *Upheavals of Thought: The Intelligence of the Emotions* (Cambridge: Cambridge University Press, 2001).

based in reality. If the evidence is inconclusive, it could be equally rational to act or wait for more information. Just as different individuals have different ways of dealing with their fears, a democratic society can choose to react in different ways to collective fears. Expert opinion about the degree to which fears correspond to the available evidence should be part of the public debate about what to do, and information about the magnitude and certainty of risks should be disseminated so that people have the opportunity to examine their fears.

In cases where evidence is not available, highly uncertain, or highly disputed, however, it may not be possible to overcome fear by deliberating about scientific evidence precisely because scientific assessments are unreliable or incomplete. The kinds of novel risks posed by dramatic new technologies expose the limitations of expert systems. When the science is so uncertain, expert analysis is more likely to reflect the background assumptions, priorities, values, and imaginations of the experts. Worse, it might reflect their self-interest or personal preferences. The challenge in these cases is not how experts can deal with the epistemic limitations of citizens but rather how citizens can deal with the epistemic limitations of expert knowledge. In cases where reason cannot guide fear, it might be rational to allow an examined fear to guide reason. In the face of a novel and serious source of danger, a sensible response is to try to remove it. A decision to ban the development of a risky new technology is an indisputably effective method for achieving this result. Uncertainty about outcomes does not justify inaction when the possibility of harm is plausible.

This emphasis on preemptive action in the face of uncertain knowledge bears affinity to the precautionary principle, but my argument differs from it in crucial respects. The precautionary principle codifies the attitude of precaution into a system of expert risk management in regulation.[33] It takes fear to be the default attitude in regard to new technologies that raise concerns about

33. For discussions of the philosophical foundations and practical uses of the precautionary principle, see Stephen Gardiner, "A Core Precautionary Principle," *Journal of Political Philosophy* 14, no. 1 (2006): 33–60; David Michaels, "Manufactured Uncertainty: Contested Science and the Protection of the Public's Health and Environment," in *Agnotology: The Making and Unmaking of Ignorance,* ed. Robert Proctor and Londa Schiebinger (Stanford, CA: Stanford University Press, 2008), 90–108; Daniel Steel, *Philosophy and the Precautionary Principle: Science, Evidence, and Environmental Policy* (Cambridge: Cambridge University Press, 2015).

the possibility of harm, but whose risks cannot be established with sufficient certainty. The principle can be interpreted either descriptively or normatively. As a descriptive claim of citizens' attitudes, it would simply be wrong; the evidence shows that most people do not uniformly fear all new technologies. Fears of risk are closely related to cultural beliefs; people fear different things depending on their value systems.[34] This suggests that citizens might be willing to take risks for some new technologies; they might well choose an attitude of hope rather than fear in the face of some unknowns. The precautionary principle cannot be justified by the claim that it tracks widespread public attitudes.

The alternative interpretation is that the principle rests on a normative judgment that fear *is* the appropriate attitude in all relevant cases of uncertain new technologies. If this is true, then it would amount to the imposition of a conservative outlook toward new technologies on a society. What might justify this? One possibility is that the historical evidence supports the conclusion that the precautionary approach is more effective overall in protecting the vulnerable or maximizing welfare.[35] This would be a persuasive defense of using the principle in areas where such evidence can be found, but it cannot justify extrapolating it to the case of new technologies where no comparable historical record exists. It may be appropriate to take a cautious attitude toward the latest pesticide in light of harms done by unregulated pesticides in the past, but this cannot be held up as a lesson for SAI. Another possibility is that the precautionary principle reflects the political victory of those who oppose new technologies over those who favor them. This may explain the cross-national differences in regulatory approaches between different countries. But of course an explanation is not a justification.

Ultimately, it is difficult to justify a blanket precautionary approach in the absence of evidence that it either responds to public attitudes or brings about the best outcomes. Since the goodness of outcomes is at least in part constituted by subjective attitudes, we arrive once again at the conclusion that the decision must be made through democratic procedures that are designed to deal with disagreements of this kind. This argument refrains from judging the appropriateness of fear versus hope as an attitude toward decisions about new

34. Dan M. Kahan, Paul Slovic, Donald Braman, and John Gastil, "Fear of Democracy: A Cultural Evaluation of Sunstein on Risk," *Harvard Law Review* 119 (2006): 1071–110.

35. Steel, *Philosophy and the Precautionary Principle*.

technologies, but stresses that the ability to act on these, and even act ahead of the evidence, must be part of democratic self-rule.

Three Objections

The most serious objection to the argument that democracies may preemptively ban certain technologies before their harms can be scientifically established is the view that it is generally better to conduct further research on a topic, and make decisions with more and better information. New research can provide more certainty about risks, and lead to the development of newer, safer, and more effective versions of a technology. Allowing further research may enhance the quality and reliability of both scientific expertise and the democratic decisions made on its basis. Since knowledge itself cannot harm, it is better to acquire as much of it as possible and regulate the potentially destructive uses of a technology in practice. This possibility cannot be ruled out; decision makers will always face the question of whether it is more advisable to wait or act preemptively. It may well be true that waiting will be the wise thing to do in many cases.

Since the objection rests on the blanket assumption that more knowledge will always enhance the quality of decision-making, however, it can be challenged by pointing out circumstances under which this is unlikely to be true. I will suggest three.

First, committing to continued research and development for an ambitious new technology can result in a society becoming locked into its deployment. If research goes on for long enough, the decision to deploy the technology may become the preferable option, not because of its intrinsic benefits, but because of the high cost of switching to a better alternative or abandoning the project altogether. The decision to continue research is not simply a neutral one in favor of more information but instead a finger on the scale in favor of deployment later on. Making a preemptive decision to stop research at an early stage can be crucial for preventing the infrastructure and institutional costs associated with lock-in at the deployment stage. This logic can be seen clearly in the governance of nuclear weapon technologies. A decision to allow research into the development of nuclear weapons makes it more difficult to govern and prevent their deployment in the future.[36] Geoengineering proposals face the

---

36. Benjamin Greene, *Eisenhower, Science Advice, and the Nuclear Test-Ban Debate, 1945–1963* (Stanford, CA: Stanford University Press, 2007).

same problem; a commitment to pursuing experiments in SAI may come at the expense of other options for mitigation and adaptation.[37]

Second, certain kinds of new knowledge can be detrimental to good decision-making in a political context, where powerful groups are likely to manipulate it to further their own interests at the expense of the common good. This will be a more serious concern if these groups are close to the source of the knowledge, such as corporations directly involved in the development of a new technology. The successful development of a new technology creates powerful interests that stand to profit from its deployment, regardless of the social costs. The problem will be particularly severe if there is uncertainty about the distribution of the costs, while the benefits will be concentrated in a few hands.[38] The dilemmas raised by the development of killer robots has this structure: those who develop the technology can expect to make large profits by selling it to interested militaries or even private actors, while it will not be clear ex ante who will be harmed the most by their potentially disastrous use in war. If the distribution of the costs and benefits takes this shape, regulatory capture will be likely. Government officials can be pressured by the powerful few to advance sectional interests rather than developing policies that serve the public good. Allowing the pursuit of certain kinds of research may thus be counterproductive to good policy making on an issue. Forcing a decision before information about the possibility of developing the technology and the distribution of harms and benefits becomes available creates a veil of ignorance that prevents the interests of the powerful from exerting a disproportionate influence over the decision.

These two responses describe conditions under which new knowledge might distort the process of democratic decision-making by changing the distribution of the expected material costs and benefits. Another response is to consider the psychological costs of waiting. Political theorist Alison McQueen observes that "a salutary and civic fear appeal must present us with an outcome that is not necessary but contingent—an outcome that it is within our power to affect."[39] At the research stage, acting on fear can be salutary in McQueen's

37. See Catriona McKinnon, "Sleepwalking into Lock-in? Avoiding Wrongs to Future People in the Governance of Solar Radiation Management Research," *Environmental Politics* 28, no. 3 (2019): 441–59.

38. Amy Sinden, "In Defense of Absolutes: Combating the Politics of Power in Environmental Law," *Iowa Law Review* 90 (2004–5): 1405–513.

39. Alison McQueen, "Salutary Fear? Hans Morgenthau and the Politics of Existential Crisis," *American Political Thought* 6, no. 1 (2017): 78–105.

sense precisely because it can produce an outcome—the removal of the fear by preventing the development of the technology. Once the technology is developed, however, the range of possible actions will be constrained by the fact that the technological change is irreversible. This alters the possible effectiveness of regulation and can engender a sense of powerlessness, as the threat of harm becomes more real and the possibilities for control move beyond the scope of existing institutions.[40]

The objection that it is always better to do more research rather than move early to prohibit it rests on a hypothesis about the likely positive effects of more knowledge on democratic decision-making. My responses are likewise hypothetical claims that focus on the reverse possibility: that more knowledge can be detrimental to good decision-making. These points can be interpreted as delineating the conditions under which a preemptive ban would be preferable. While seeking more information can be a good rule of thumb, there will be exceptions to the rule. I have suggested that preemptive restrictions can be effective and empowering political acts in cases that appear likely to lead to technological lock-in, regulatory capture, or inaction with the passing of time. The argument ultimately depends on ascertaining when and whether these hypotheses hold; it would therefore benefit from empirical research into the efficacy of moratoriums and prohibition as regulatory tools under different circumstances.

A second objection is that allowing collective fears to play a role in the regulation of technology encourages a politics of fear appeals.[41] In particular, it raises the worry that citizens will be susceptible to manipulation. Individuals who are closely involved with the research and development of a technology hold great power to shape the collective imaginary about its potential impacts. Those who possess the ability to shape people's fears and hopes of the future also hold the power to direct their behavior, and may use this power in self-serving ways.

The objection can be analyzed in two parts. The first is the worry that public discourse will be guided by appeals to the imagination.[42] Since scientific

---

40. For a discussion of problems whose governance becomes "beyond imagination," see Nancy L. Rosenblum, "Governing beyond Imagination: The 'World Historical' Sources of Democratic Dysfunction," *Boston University Law Review* 94, (2014): 649–69. Rosenblum gives the examples of climate change and surveillance technologies.

41. On the attraction and pitfalls of a politics of fear appeals, see, for example, McQueen, "Salutary Fear?"

42. Sheila Jasanoff and Sang-Hyun Kim develop the concept of "sociotechnical imaginaries" to describe this. See Sheila Jasanoff and Sang-Hyun Kim, "Containing the Atom: Sociotechnical Imaginaries and Nuclear Power in the United States and South Korea," *Minerva* 47, no. 2 (2009):

knowledge about the impacts of a new technology is highly uncertain, political debate around its promises and dangers will inevitably be speculative. This is not unique to the context of technology; political claims about the distant future always engage the imagination, and good politicians understand that their ability to create a persuasive narrative for the future, whether one of hope or fear, is crucial for winning support for their policies. This strikes a contrast with antipolitical, expert-led approaches that center on controlling future outcomes by quantifying and weighing risks guided by a principle of harm prevention. I have argued that these two approaches are both useful and should be complementary in the governance of new technologies. The joint requirement of a harm principle and consent principle ensures this. Still, the limitations of expert risk management in cases of indeterminacy makes it inevitable that speculative claims will play a larger role. This is not intrinsically problematic as long as it is possible for citizens and decision makers to critically examine the plausibility and attractiveness of competing narratives.

The crux of the objection is the second part: the worry that the development of technologies necessarily gives elites—scientists, tech developers, and Silicon Valley billionaires—the power to guide the public imaginary, and could easily lead to its manipulation. While the campaign for a ban on killer robots was an example of elite-led alarmism, in other cases, scientists have attempted to present geoengineering schemes as the key to a utopian future, where both environmental and distributional problems would be resolved by easy and cheap technological interventions that cut across political divisions.[43] The demographic composition of those who shape the public imaginary and set the agenda is a further source of concern. An analysis of news coverage on geoengineering revealed that 97 percent of public assertions on the topic were made by male researchers, and only 3 percent by women.[44]

119–46; Sheila Jasanoff and Sang-Hyun Kim, eds., *Dreamscapes of Modernity: Sociotechnical Imaginaries and the Fabrication of Power* (Chicago: University of Chicago Press, 2015).

43. See, for example, Oliver Morton, *The Planet Remade: How Geoengineering Could Change the World* (Princeton, NJ: Princeton University Press, 2016). For a study of the use of equity justifications in visions of geoengineering, see Jane Flegal and Aarti Gupta, "Evoking Equity as a Rationale for Solar Geoengineering Research? Scrutinizing Emerging Expert Visions of Equality," *International Environmental Agreements: Politics, Law and Economics* 18, no. 1 (2018): 45–61.

44. Holly Jean Buck, "Climate Engineering: Spectacle, Tragedy or Solution? A Content Analysis of News Media Framing," in *Interpretive Approaches to Global Climate Governance: (De)constructing the Greenhouse*, ed. Chris Methmann, Delf Rothe, and Benjamin Stephan (New

The worry about an expert-dominated public discourse is a serious one, and throughout this book, I have offered many ways to mitigate its effects. The domain of technology is also governed by expert knowledge that is often impenetrable to nonexperts, who may find themselves confronted with a choice between rival visions of the future that may seem equally fantastic and alien. As I have been arguing in this book, though, the response to this problem should not be to remove the decision from democratic participation but rather to think about institutional arrangements that would facilitate public participation and allow citizens the chance to scrutinize competing technological visions. The science court proposed in chapter 4 could play a part in the regulation of new technologies by allowing ordinary citizens to examine the assumptions, limitations, and uncertainty of different expert views about the harms and benefits of new technologies.[45] This would increase the power of citizens, and make them less susceptible to the dystopian or utopian appeals of experts and other elites.

A final objection concerns the efficacy of a democratic decision to restrict research in one country when scientific research is increasingly international. Even if scientists in one country halted research, those in other ones could continue to pursue it. This would lead to one of two outcomes. One possibility is that the findings would be shared globally, thus allowing governments or corporations to use the technology, and bringing about the feared outcomes that had given rise to restrictions in the first place. The restrictions would be useless. Another, possibly worse scenario is that countries where scientists are allowed to pursue dangerous science would keep the findings to themselves, and use them in ways that harm or disadvantage other countries, such that any country with restrictions on the research would be worse off for lacking the relevant scientific knowledge to defend its interests or keep up with the competition. This would also make it more difficult for scientists in these countries to hold other scientists accountable for accidents or misuses of the technology.

The international dimension of the problem is more significant in some areas than in others. Some areas of research are pursued by a small number of researchers in a single country. In the case of SAI, only one team in the United States is

---

York: Routledge, 2013), 166–83. See also Holly Jean Buck, Andrea R. Gammon, and Christopher J. Preston, "Gender and Geoengineering," *Hypatia* 29, no. 3 (2014): 651–69.

45. This bears some resemblance to Thomas Hobbes's recommendation that people should test those making apocalyptic claims in order to expose them as false prophets. See Alison McQueen, *Political Realism in Apocalyptic Times* (Cambridge: Cambridge University Press, 2017), 125–27.

currently close to the stage where running large-scale field experiments is a real possibility.[46] One researcher estimated that there are less than a hundred researchers in the world working on the topic, and most are only running models.[47] A restriction on field experiments by the US government today could halt research in this area for some time. Democratic restrictions will obviously be more effective in cases like this, where the research is unlikely to go ahead in another country due to constraints on resources, knowledge, or interest. If countries at more advanced stages of the research process place restrictions first, scientists in other countries might be less willing to enter the field later on because of the diminished opportunities for knowledge sharing, collaboration, and exchange. Finally, restrictions in one country might always set an example that persuades others to follow suit. The domestic policies of a few countries at the forefront of research often have outsized effects on the rest of the world. There will always be some areas of scientific research, however, that cannot be effectively restricted except through international agreements given the strong competitive advantages of pursuing the research even if, or especially if, other countries are not doing so. Killer robots fall in this category. In these cases, societies concerned about the dangers of research must push for an international ban on their development, following the example of the robotics researchers discussed earlier.

Striking a Balance

The ethical guidelines for regulating research with human subjects aim to strike a balance between protecting the welfare of subjects and respecting their autonomy. The former task is entrusted to experts, while the latter requires allowing individual choice. The application of these principles to the regulation of high-risk new technologies can, in turn, be interpreted as an effort to find the proper balance between expert-driven risk analysis and democratic decision-making. I have outlined a role for both expertise and democratic decision-making in the regulation of new technologies at the research stage. But my two main arguments—that research may be restricted on the basis of foreseeable harms from the application of a technology and subjective determinants of risk such as fear—tilt the balance in favor of social and political concerns.

46. Jeff Tollefson, "First Sun-Dimming Experiment Will Test a Way to Cool the Earth," *Nature* 563, no. 7733 (2018): 613–15.

47. Sarah Sax, "That Geoengineering Gender Problem," *Canada's National Observer*, December 23, 2019, https://www.nationalobserver.com/2019/12/23/news/geoengineering-gender-problem.

7

A Political Theory for
Uncertain Times

TWENTIETH-CENTURY REFLECTIONS on politics and expertise were structured by a Weberian paradigm that rested on two foundational claims. The first
was that science and instrumental rationality had made the world transparent
and predictable, rendering it amenable to human control based on objective
scientific calculations. The second was that these scientific calculations could
not deliver answers about their own meaning, significance, and purpose. Since
science could not force choice, people were forced to be free to choose.[1] Weber
assigned responsibility for this choice to a charismatic political leader, while
Habermas offered a democratic reworking of this model by arguing that a ratio-
nally deliberating public should determine the ends toward which scientific
inquiry should be directed.

The Weberian picture was always meant to be a regulative ideal. Weber
recognized that bureaucracies were partial and politicized in practice. He insisted that they must be kept out of politics precisely because he recognized
the impossibility of complete success.[2] There are consequences, however, to
poorly approximating an ideal that cannot be attained. The possibility of neutral and reliable expertise has defined theories about the relationship between
experts and politics, while the myriad ways in which experts fall short of the
ideal in practice have not fundamentally altered our thinking about the problem. From advisory institutions that entrust experts with the factual basis of

---

1. Sheldon Wolin, "Max Weber: Legitimation, Method, and the Politics of Theory," *Political
Theory* 9, no. 3 (1981): 401–24.

2. Jennifer M. Hudson, "The Bureaucratic Mentality in Democratic Theory and Contemporary Democracy" (PhD diss., Columbia University, 2016).

decisions to deliberative experiments that expect citizens to deliberate about ends in light of the facts provided by experts, to normative theories of democracy that maintain that ordinary citizens need not and should not be involved with the technical aspects of government and administration, the influence of this model is still widespread in both the theory and practice of democracy.

This book has argued for a different approach that starts from a philosophically precise account of the ways in which the knowledge of scientists is not adequate to the immense political task of neutrally but reliably informing political deliberation about ends. Chapters 2 and 3 examined how science goes beyond this role in some ways, and falls short of it in others. It goes beyond its role because scientific methods and theories incorporate judgments about the significance and meaning of different kinds of knowledge, often derived from background beliefs about appropriate scientific or political ends to pursue. Decisions about what kinds of knowledge to seek and how, frame these debates and set the agenda. While these considerations should be part of a democratic deliberation about ends, scientists implicitly or explicitly make these judgments at earlier stages of research, from the selection and funding of research agendas, to the formation of concepts, selection of methods, design of experiments, and construction of models. The science we have thus ends up being more compatible with some ends over others in ways that are difficult to detect.

At the same time, science falls short of fulfilling the role that is expected of it because it is uncertain, incomplete, and subject to disagreement. A model in which scientists can be counted on to provide certain information to policy makers about how to attain a specified end is therefore unrealistic except in a narrow class of simple cases. Under conditions of uncertainty and ignorance, which define the majority of policy questions requiring scientific advice, the facts and the purposes for which they will be used must be determined jointly. When scientific advisers make judgments that combine factual and evaluative considerations in order to provide more useful advice, however, they jeopardize the neutrality that legitimates their authority in a democratic system. The standard ideal for the relationship between epistemic and practical authority is unstable.

In its place, I proposed a model that emphasizes ongoing democratic scrutiny and input on scientific issues along with institutional innovations designed to facilitate this from the earliest stages of research, thus blurring the boundaries of the division of labor between scientists, politicians, and ordinary citizens. I argued that policy-relevant scientific claims must be examined democratically to expose their uncertainty and limits, and that the basis of policies must be debated and accepted collectively—for instance, by citizen juries in a science court—with more awareness of the strength, bias, and implications of the findings. Since the framing of issues and possible alternatives are usually determined by the available research, I also argued that the institutions that fund and govern science must be responsive to democratic priorities. I defended the necessity of direct democratic input into decisions to fund or defund science as well as indirect democratic innovations such as randomization in funding, more diverse funding committees, and stipulations to fund dissenting views.

The language of democratizing science is used so frequently, going at least as far back as Dewey, that I hesitate in claiming it, although the ideas that I have defended here can be captured by it. I should stress, though, that there is little continuity in how this term has been used by different thinkers. It has variously meant modeling democracy after science as a community of inquiry, improving the public understanding of science, opening up the laboratory to contributions by citizen scientists, incorporating local or lay knowledge in scientific research, increasing transparency and data sharing in science, and even organizing the internal governance of the scientific community more democratically. In most cases, the main goal has been to improve the use of science in policy decisions, but without granting citizens any meaningful control over the agenda and activities of science itself.[3] The widespread practice of treating science for policy and policy for science as two quite separate subjects is yet another sign of this.

In contrast, I have argued that truly democratizing science means both opening up scientific claims to public scrutiny, and ensuring more democratic input into the governance of science and technology, focusing especially on decisions about what knowledge should be produced. It is possible to imagine a society in which science would be conducted fully autonomously from social concerns and funded only for its intrinsic value. Perhaps there would be nothing wrong with such a society, but its attitudes toward science would be markedly different from those of most modern societies. Since modern states are committed to relying on scientific knowledge for policy purposes—at least in principle, if inconsistently in practice—and many have justified spending large

3. This point is also made in Philip Mirowski, "The Future(s) of Open Science," *Social Studies of Science* 48, no. 2 (2018):171–203. For a notable exception to this rule, see Philip Kitcher, *Science, Truth, and Democracy* (Oxford: Oxford University Press, 2001).

amounts of public funds on science on the basis of the public benefits of doing so, the political impact and influence of science cannot be ignored in decisions regulating earlier stages of the practice.

Trust

The more skeptical and adversarial approach toward science advocated in this book will inevitably raise some eyebrows at a time when distrust and denial of science are on the rise.[4] In response, I will appeal to Jeremy Bentham's views on the appropriateness of trust and distrust in politics. Bentham maintained that trust is properly placed in institutions that systematize an attitude of healthy distrust. He believed that such carefully designed institutions would, in turn, facilitate good judgments about which particular people or claims could be trusted. His crucial point was that trust in politics should be earned by actors within such a carefully designed system rather than encouraged wholesale without regard to the differences that make some individuals more trustworthy than others.[5] This summarizes my approach to trust.

Trying to increase an undiscriminating trust in science and scientists without regard to the content of their claims would betray the spirit of science itself. The healthy distrust embodied in a system intended to question and challenge expert claims at various junctures—specifically expressed in the design of the science court—is important for ensuring the adequacy of science for furthering democratic purposes.[6] In turn, institutions and practices designed to facilitate good judgment and careful evaluation of scientists would be more likely to prevent undiscriminating denial, which after all is the mirror image

---

4. There is some evidence challenging this prevalent narrative of the distrust in science. A recent study by the American Academy of Arts and Sciences finds that trust in the leaders of science has stayed constant in the United States over the past five decades. See *Perceptions of Science in America: A Report from the Public Face of Science Initiative* (Cambridge, MA: American Academy of Arts and Sciences, 2018).

5. Jeremy Bentham, *Securities against Misrule and Other Constitutional Writings for Tripoli and Greece*, ed. Philip Schofield (Oxford: Clarendon Press, 1990). I follow the interpretation in Jonathan Bruno, "Vigilance and Confidence: Jeremy Bentham, Publicity, and the Dialectic of Political Trust and Distrust," *American Political Science Review* 111, no. 2 (2017): 295–307.

6. For an account of the differences between healthy distrust and conspiracy thinking, see Nancy L. Rosenblum and Russell Muirhead, *A Lot of People Are Saying: The New Conspiracism and the Assault on Democracy* (Princeton, NJ: Princeton University Press, 2019).

of undiscriminating trust. The challenge is not to determine in the abstract how much trust is appropriate for science but rather to design good institutions for evaluating science, which will earn trust themselves and facilitate trust in the scientists who participate in them.

This book enhances our understanding of the dynamics of public trust and distrust in science by changing what should be seen as the correct institutional locus of trust. Science is usually thought to merit public trust because of the way its internal methods and procedures ensure the reliability of findings. Peer review, replication, transparency, and shared norms of doubt and skepticism are all intended to detect and correct potential mistakes, and improve knowledge through intersubjective criticism. The failure of these scientific procedures and norms then becomes the obvious explanation for the rise of the public distrust in science. The rise of research biased by financial interests, the exposure of fabricated data, replication crises, and increased retractions from journals are frequently offered among the reasons for public mistrust.

But while the internal procedures of science are no doubt crucial for ensuring the reliability of scientific claims as truth claims, my argument suggests that they are not sufficient for ensuring the reliability of science for democratic use. Additional considerations emerge at the stage of use: the appropriateness of relying on findings in a particular context, their adequacy for a specific purpose, the consequences and distribution of mistakes, the significance of different kinds of knowledge, the role of important omissions in knowledge, the dangers of misuse, and so on. These are just as critical in determining the reliability of science for a particular purpose as the evidentiary support enjoyed by a hypothesis. Even if the scientific process has worked as it should, and fraud or deception are absent, there are still good reasons to question the use of particular scientific findings in a political context. I have identified this as a crucial and distinct problem with the use of science, and argued that we should pay more attention to the question of who makes the decisions about significance, adequacy, and relevance as well as examining how these are shaped by the way that inquiry has been conducted in the first place. A science that is more responsive to democratic concerns in the ways that I have specified is likely to earn greater trust too.[7]

---

7. For a nuanced argument on the public trust in science from within Kitcher's well-ordered science framework, see Gürol Irzık and Faik Kurtulmuş, "Well-Ordered Science and Public Trust in Science," *Synthese* (2018), doi.org/10.1007/s11229-018-02022-7.

Communicating Science

A common approach to addressing problems at the intersection of science and politics is to view them as failures in the communication of science to the public. A large literature focuses on developing strategies to improve the effectiveness of communication in order to increase public understanding and therefore public trust. I argued that this approach is insufficient on its own because it assumes that problems around science would be solved if only ordinary citizens could be made to understand scientists better. It also expects citizens to accept science as it is presented to them, without engaging with the reasons why they may be right not to do so. While communication is undoubtedly an important aspect of the issue, I think there are other dimensions that are more fundamental. Starting and ending with communication is not enough.

That said, the argument in this book does have implications for how science should be communicated. A nuanced account of the difficulties in the relationship between science and democracy is a crucial step for thinking about good communication. With such an account in hand, we can reflect more meaningfully on the communication strategies appropriate for interactions between scientists and the public. My emphasis on the importance of identifying the limitations of scientific knowledge implies that scientific findings should be communicated in ways that emphasize their uncertainty, incompleteness, and potential biases instead of trying to conceal them.[8] Strategies that try to persuade the audience of the scientist's position at all costs would be misguided. I proposed the use of adversarial structures to address this problem; these would function more effectively if supplemented with role responsibilities for participating scientists. Both the science court and broader deliberative processes in the public sphere would work better if scientists shared their assessment of the uncertainties and gaps in their research, instead of trying to make findings appear as certain and seamless as possible.

My argument supports the view that the proper rhetorical approach to communicating science should be one that respects the audience's agency, and tries to enhance it, as opposed to merely attempting to instruct or persuade.[9] Scientists in the public sphere or policy-making circles should try to help their

---

8. The East Anglia climate scientists' email debacle—"Climategate"—is a cautionary tale of what can happen when they don't.

9. Michael Lamb and Melissa Lane, "Aristotle on the Ethics of Communicating Climate Change," in *Climate Justice in a Non-Ideal World*, ed. Clare Heyward and Dominic Rose (Oxford: Oxford University Press, 2016), 229–55.

audience make its own judgments instead of imposing scientific judgments on the audience. Rhetorical moves that demonstrate the speaker's willingness to share power with the audience and submit claims to the audience's scrutiny can facilitate this. Reflections on the appropriate forms of communication should go beyond considerations about effectiveness and focus on the right ethical stance of the speaker vis-à-vis the audience. Success in communication should not be measured by the ultimate agreement of the audience with the speaker but rather by how empowered members of the audience become in making up their own minds on the issue.

The Role of Scientists in Public Life

This book assigns an important public role to scientists. The science court would not take off if scientists refused to participate. This is a demanding role and requires quite different skills than the ones associated with being a good scientist. Obviously, not all scientists must become active in public life, as long as a sufficient number are willing to do so. Debating and cross-examining an adversary in front of a public audience, as the science court would have it, can be exhausting, and may require heroic patience and goodwill from scientists, but I hope that I have provided sufficient reasons to believe that the democratic value of these efforts would be great.

One crucial obstacle to a more prominent public role for scientists is that public engagement is usually not rewarded by career advancement and may even be punished. This could lead to selection bias in the kinds of scientists that accept public appearances. There are currently many examples of top scientists in areas ranging from epidemiology and nuclear physics to climate change and genetics who choose to become public figures even under highly adverse circumstances. This suggests that we need not despair. Still, public participation by scientists could be further encouraged through professional incentives. Counting public engagement as a plus or even requirement in promotion would be one way to improve the alignment of scientists' professional priorities with increased public engagement.

A Note on Uncertainty

I think it is fitting to end with a note on uncertainty given its central role in both scientific inquiry and the argumentative structure of this book. Each chapter can be read as an attempt to expose the challenges of balancing our need for knowledge with an awareness of the inherent uncertainty and incompleteness of the knowledge available. Chapter 2 explored how scientists' judgments about significance in the face of uncertainty and the impossibility of attaining complete knowledge results in findings that are skewed toward the pursuit of certain ends over others. I pointed out how the transition from lab to use creates additional uncertainties that scientists are not well placed to deal with. Chapter 3 showed how the need for judgment under uncertainty poses difficulties for the simple division of labor between scientific and democratic authority, and underscored the futility of scientific advisers trying to settle uncertainty through prolonged debate. Chapter 5 exposed the uncertainty underlying the claim that the activities of a highly autonomous scientific community would create knowledge that is significant and useful from a democratic perspective. Finally, chapter 6 pointed out the difficulties of trying to assess expected benefits and harms from research under uncertainty, and argued that collective emotions such as fear and hope become more salient for democratic decision-making when experts cannot offer good predictions. I concluded that we should not assume that a society is always benefited by unrestricted inquiry.

There are two problematic ways of responding to uncertainty and a lack of knowledge: the first is to see it as inevitably leading to procrastination and inaction, and the second is to view it as a state that must be overcome through more research or deliberation. Corporations have strategically used the first approach to delay action on climate change and smoking, while scientists have often unwittingly contributed to delays by taking the second route. I argued by contrast that the inevitability of uncertainty and the limits of knowledge should not be excuses for inaction but rather prompts for rethinking how decisions involving science should be made, who should make them, and through which procedures. Ultimately, decisions must always be made about what to accept as useful knowledge for a particular purpose, and when to act on it under ongoing uncertainty. There are no right answers for these, and we should not expect science to answer them for us. While the scientific community may have its own methods for settling truth claims, decisions about the significance, usefulness, and adequacy of knowledge are never purely scientific; they are deeply political too. It is only appropriate, then, that they should be made through political processes of scrutiny, debate, and judgment.

8

Epilogue: COVID-19

THE COVID-19 PANDEMIC, which started in Wuhan, China, in December 2019 and spread rapidly throughout the world, put science and scientists under a spotlight like rarely before. This was truly a life-and-death matter on which governments around the world depended on scientific advice. Previously unheard-of scientific advisory groups became household names, and scientific advisers appeared on daily press conferences alongside prime ministers and presidents. Much can be said about the merits of different responses to the crisis and the mistakes that contributed to the death toll. Physicians, public health experts, and social scientists are already studying the effects of COVID-19 policies, and will continue to do so for years to come. My aim in this epilogue is not to contribute to these first-order discussions about policies and outcomes but instead to examine the second-order question of what the pandemic revealed about decision procedures at the intersection of science and politics.

The pandemic hit just as I was finishing this book. This seemed to be awkward timing at first, but I soon realized that it was a unique opportunity to test my ideas on a case of unprecedented significance and impact. I did not have the benefit of the COVID-19 experience while developing my arguments, but having written the book, I could interpret the challenges of COVID-19 through my theoretical framework. The complex scientific and political dynamics of the COVID-19 pandemic, in turn, allowed me to refine the details of my arguments, and add nuances and caveats. In the following, I will show how the questions addressed in each chapter of the book—from the role of values and uncertainty in science to the paradox of scientific advice, from the need for public participation to the role of democratic input into decisions to fund science—became salient in the COVID-19 context. I hope this will demonstrate the critical and clarificatory power of my arguments for making sense of this episode while also providing concrete illustrations of ideas discussed more abstractly earlier in the book.

The COVID-19 pandemic is a case that fits the scope conditions outlined for this book almost perfectly. It is a high-stakes, high-uncertainty issue, where evidence was scant, the science not settled, and political decisions were urgent. Scientists still have no answers on many key scientific questions. This ongoing uncertainty is uniquely valuable for thinking about the experience of making important decisions without knowing what is right, scientifically or morally. While hindsight will make it easier to draw general and reliable conclusions, full possession of the answers might make it more difficult to appreciate the challenges that scientists, politicians, and the public faced in the midst of the crisis. This epilogue is an attempt to render this experience as it unfolds.

Following the Science

The discussion in chapter 2 on the role of values and purposes in scientific models was particularly relevant during the early months of COVID-19, when government responses relied heavily on a small number of models. As I pointed out earlier, models cannot be empirically verified because they rest on assumptions and approximations that are strictly false. They merely produce predictions that are more or less adequate for particular purposes. The usefulness of a model depends on the match between the purposes and needs assumed by modelers, and the purposes and needs of policy makers and the public. Yet if there is an urgent need for action and only a small number of models available, crucial policy decisions will be influenced by scientists' assumptions. After all, it is better to rely on available models than not. The early COVID-19 models thus played a critical role in shaping government responses as well as determining their limitations.

Around mid-March, the UK and US governments shifted their more relaxed pandemic response suddenly and dramatically in reaction to an extreme forecast from the Imperial College modeling group, which predicted 500,000 deaths in the United Kingdom and 2.2 million in the United States. At around the same time, the University of Washington's Institute for Health Metrics and Evaluation (IHME) model was still predicting around 200,000 deaths in the United States.[1] These wildly differing predictions led to criticisms that the modeling enterprise had largely failed. The divergence in predictions was due

1. Christopher Avery, William Bossert, Adam Clark, Glenn Ellison, and Sara Fisher, "An Economist's Guide to Epidemiology Models of Infectious Disease," *Journal of Economic Perspectives* 34, no. 4 (Fall 2020): 79–104.

in part to the fact that modelers had to make many speculative and often ad hoc assumptions on key parameters, such as infection rates, infection fatality ratios, immunity levels, and asymptomatic transmission. These naturally affected the accuracy of the predictions.[2] But just as important, these two models were intended for different purposes, and these purposes influenced the assumptions and mathematical techniques that they used. In turn, these assumptions and techniques determined the strengths and weaknesses of the models as well as the direction of their errors and biases. The lack of public scrutiny of the purposes and assumptions of different models exacerbated the intrinsic problems due to uncertainty and incomplete knowledge.

The Imperial College modelers intended to produce forecasts for eleven possible policy responses by running disease transmission simulations under varying assumptions.[3] The forecast that made headlines and led to the changes in government policy was their most extreme scenario. It assumed no policy response and no change in individual behavior. This was one of the least plausible possibilities, but it garnered the most attention. The media did not always report the fact that this forecast was based on the assumption of no government response.[4] The IHME model used a different approach because its purpose was different: it aimed to provide a short-term death count that would help hospitals plan for hospital bed and intensive care unit demand rather than supplying a detailed simulation of the progress of the disease over time.[5] The modelers simply fit a curve to observed death rates from the first months of the pandemic. They assumed that the death rate would be a bell-shaped curve, and used the existing data from China and Italy to find its parameters. This curve-fitting approach had inherently limited efficacy for modeling the long run and was not responsive to variations in mitigation policies across regions.

More public debate about the assumptions and purposes of the models could have prevented some of the misguided accusations directed at the modelers. Perhaps more important, it could have exposed the fact that both models shared critical limitations, including ones due to questionable judgments

2. Sibel Eker, "Validity and Usefulness of COVID-19 Models," *Humanities and Social Sciences Communications* 7, no. 54 (2020): 1–5; John P. A. Ioannidis, Sally Cripps, and Martin A. Tanner, "Forecasting for COVID-19 Has Failed," *International Journal of Forecasting* (forthcoming).

3. Eker, "Validity and Usefulness of COVID-19 Models."

4. Avery et al., "An Economist's Guide to Epidemiology Models."

5. Eker, "Validity and Usefulness of COVID-19 Models."

about what was significant and what could be bracketed. For instance, neither the Imperial College nor IHME model considered how infection rates would change based on social behavior. It was only later that more nuanced models incorporated assumptions about individual behavior under different government policies, including assumptions of heterogeneous behavior across different groups.[6] Some economists hypothesized that contact rates among individuals would likely be endogenous to infection rates, with people following restrictions more carefully as infections went up and vice versa.[7] Efforts to increase verisimilitude create their own difficulties, though, as behavioral assumptions are even more speculative and value laden. Both the contact rate and its relationship to infection rates depend in complicated ways on variables like class, age, and ethnicity that influence social interaction patterns, such as through the effects of housing conditions, occupational requirements, and cultural norms.

Models also made questionable assumptions and important omissions about outcomes of interest, concept definitions, and relevant policy scenarios. For instance, both the Imperial College and IHME models focused on the total death count rather than using more nuanced health measures such as quality-adjusted life years or studying health effects across population subgroups.[8] Moreover, they studied short-term health outcomes, and entirely neglected the economic and social impacts of policies. This meant that they failed to take a holistic approach to health outcomes overall, and left out the mental and physical health toll of social isolation and a severe economic downturn, increased domestic violence and substance abuse rates, delayed treatments for other diseases, and missed vaccination schedules for children. The policy responses included in the simulation depended on views about what would be acceptable ethically and politically too. For example, some UK advisers initially dismissed the possibility of a strict lockdown on the grounds that it would be politically unthinkable.[9]

6. Avery et al., "An Economist's Guide to Epidemiology Models."

7. Flavio Toxvaerd, "Equilibrium Social Distancing," Cambridge-INET Working Paper Series No: 2020/08, March 2020, http://www.econ.cam.ac.uk/research-files/repec/cam/pdf/cwpe2021.pdf; Jussi Keppo, Marianna Kudlyak, Elena Quercioli, Lones Smith, and Andrea Wilson, "For Whom the Bell Tolls: Avoidance Behavior at Breakout in COVID19," technical report, working paper, 2020.

8. Ioannidis, Cripps, and Tanner, "Forecasting for COVID-19 Has Failed."

9. Stephen Grey and Andrew MacAskill, "Special Report: Johnson Listened to His Scientists about Coronavirus—but They Were Slow to Sound the Alarm," Reuters, April 7, 2020.

I do not mean to suggest that particular choices, assumptions, or omissions were unjustified. Increasing the complexity of models could make predictions useless, and assumptions about longer-term, behavioral, and economic effects are likely to be even more uncertain and contested than assumptions about disease transmission. Nevertheless, these examples illustrate my argument that value-based scientific choices constrain political action in ways that can be difficult to detect and challenge. They also underscore how crucial it is to examine how these modeling choices shape predictions, and how the uncertainty and errors of different models depend on controversial assumptions about which variables are significant, which policies ought to be considered, or what constitutes a good measure of health. When governments take model forecasts seriously, the assumptions and mistakes of modelers end up having serious real-world consequences. These difficulties are more acute if there are only a small number of models and they have similar biases, which was the case in the earlier stages of the crisis.

A more general conclusion that I want to draw is that it would be valuable to understand how scientists actually make these modeling choices under uncertainty. An ethnography of the disease modeling community could be helpful in revealing how scientists make innumerable small yet consequential modeling choices and which resources they use in the face of large gaps in knowledge.[10] Do scientists make precautionary assumptions or try to aim for averages? How do they even decide what counts as precautionary? Do they make decisions individually or together? To what extent are these based on tacit scientific knowledge, and what extent on personal moral and social beliefs? Exploring the combination of social and scientific factors that determine modeling choices, and tracing their impact on the resulting models, would be a promising avenue for future research.

Citizens with different political views will naturally favor different policy responses to a pandemic. Some will prioritize the economy, while others will put health and safety first, and still others will insist that a certain understanding of freedom puts hard constraints on which policies the government may adopt. These conflicts must be resolved politically. The problem is that models themselves incorporate assumptions on these points, thereby preempting and circumscribing political debate, and ruling out some options entirely. The

---

10. For the role of ethnography in political theory, see Matthew Longo and Bernardo Zacka, "Political Theory in an Ethnographic Key," *American Political Science Review* 113, no. 4 (2019): 1066–70.

challenges of using models during COVID-19 thus reinforced a central message of this book: it is crucial for nonscientists to scrutinize the assumptions
that scientists make, understand the uncertainty of their claims, and be clear
about the issues on which they are completely ignorant.

How did processes of scrutiny and criticism play out during the pandemic?
There were many good examples of journalists, bloggers, data scientists, and
others simplifying, clarifying, and visualizing complex scientific information
for the public. One article explaining how to fight the virus by alternating
between periods of lockdowns and cautious opening up—titled "Coronavirus: The Hammer and the Dance"—was viewed over sixty million times and
translated into dozens of languages.[11] By contrast, official scientific advisory
processes fell short in adopting public-facing approaches that would allow real
scrutiny and accountability. Even while scientific advisers appeared frequently
on press conferences with politicians, the advice that they actually gave governments was kept secret. This made it difficult for citizens to assess whether
their interests were being represented, whose interests were prioritized, and
whether any reasoned trade-offs were being made at all.

The secrecy also made it impossible to determine how scientific advisers
struck the balance between neutrality and usefulness—the dilemma discussed
in chapter 3. There is currently not enough information to determine whether
scientific advice was guided by scientists' own moral judgment, the political
commitments of their governments, or some other conception of the public
interest. Yet there were a few striking examples of US scientists decisively moving away from the norm of neutrality in the public sphere. In one case, scientists who had been warning against the risk of infection at religious gatherings
and antilockdown protests later came out in support of the protests against
systemic racism and police brutality in the wake of George Floyd's killing by
the police in Minneapolis. While the COVID-19 risks of these gatherings were
comparable, scientists' belief in the rightness and urgency of the Black Lives
Matter protests led them to maintain that the additional risks were justified
from a public health perspective.[12] Hundreds of epidemiologists and public
health professionals signed a letter arguing that the public health messaging
for antilockdown and antiracist protests should not be the same. They declared

11. Tomas Pueyo, "Coronavirus: The Hammer and the Dance," March 19, 2020, https://
tomaspueyo.medium.com/coronavirus-the-hammer-and-the-dance-be9337092b56.

12. Dan Diamond, "Suddenly, Social Justice Matters More Than Social Distance," *Politico*,
April 6, 2020.

that antiracist protests "must be supported" even if they increased infections, but a similarly permissive attitude was not required for antilockdown protests.[13] Unsurprisingly, Republicans disagreed with this judgment, and accused scientists of liberal bias and double standards.

In chapter 3, I developed a qualified argument against the ideal of neutrality in scientific advice. I emphasized that value judgments are crucial for making advice relevant and useful, but I also warned that it is essential for different values to be represented in public advice and that this is extremely difficult. Since a small body of scientists can never represent all citizens, I argued that their judgments must be submitted to broader scrutiny and challenge, especially from those with different value commitments. The public health messaging during the protests highlighted an important problem on this score: in an overwhelmingly liberal scientific community, it was difficult for alternative political viewpoints to be adequately represented in scientific advice. While politicians and journalists noted and criticized scientists' inconsistent messages, not much disagreement came from within the scientific community. I argued in chapter 3 that organizing dissenting views from scientists themselves toward a public audience could have a crucial epistemic and legitimating role in the public uptake of science. Such dissent was missing from this episode.

Meanwhile, in the lead-up to the 2020 presidential election, many scientists took unusually open political stances against the Trump administration's dismissal of scientific evidence, and its open hostility toward science and scientists. Eighty-one Nobel Prize winners signed a letter supporting the candidacy of Joseph Biden. The journal *Nature* endorsed Biden too, while the *New England Journal of Medicine* wrote a harsh editorial criticizing the incompetence of the administration—a first in its 208-year history. These were necessary defenses against the administration's political attacks on science, but they may have reinforced the perception that the scientific community favors liberal values. A national public opinion survey conducted during the pandemic showed that many citizens were unsure about the extent to which the scientific community represented their values.[14] A majority of the participants agreed

13. "Open Letter Advocating for an Anti-Racist Public Health Response," June 5, 2020, https://www.calvoices.org/post/open-letter-advocating-for-an-anti-racist-public-health -response-to-demonstrations.

14. John H. Evans and Eszter Hargittai, "Who Doesn't Trust Fauci?" The Public's Belief in the Expertise and Shared Values of Scientists in the COVID-19 Pandemic," *Socius* 6 (2020), doi.org/10.1177/2378023120947337.

with the claim that scientists understood the spread of the virus and ought to have a fair amount of influence in handling the pandemic. But many were ambivalent about the claim that scientists' values would be consistent with theirs on a life-and-death matter. "Neither agree nor disagree" was the most commonly selected response, and Republicans and independents were much more likely than Democrats to disagree.

This pattern is troubling for theories that maintain that scientists, whether in their role as advisers or researchers, ought to represent the values and interests of the public. If the scientific community is not descriptively representative of the rest of the public, and if most citizens do not trust scientists to adequately represent their values on a life-and-death matter, arguments based on representation face serious obstacles. By contrast, these findings are not troubling for my argument because I maintain throughout this book that achieving democratic representation within and through scientific research and advice is a tall order. I stress the need to open up scientific claims to broader scrutiny in part for this reason. The observation that most scientists support one candidate or political party over another supports the view that it is better to assume that scientists are not representative of the broader community when we theorize the relationship between science and democracy. Still, I also argue that democracies have a stake in ensuring a diversity of scientific viewpoints within the scientific community on the grounds that this facilitates public scrutiny and mitigates the political influence of any particular set of value judgments. While it may not be intrinsically problematic that most scientists align with one political party, especially if their scientific claims are truly open to scrutiny and dissent, it could become problematic if partisan alignment skews the content of scientific research and advice in ways that consistently favor the interests of the same group of people. More empirical research is necessary to examine the links between partisan affiliation and scientific outputs.[15]

Let me set aside issues of partisanship now and turn to the role of scientific disagreement during COVID-19 more broadly. One of the main arguments of this book is that organized forms of scientific dissent that are directed toward a public audience can play a crucial democratic role. Chapter 3 explored the possibility of written dissents from scientific advisory committees, and

---

15. See, for example, Eitan D. Hersh and Matthew N. Goldenberg, "Democratic and Republican Physicians Provide Different Care on Politicized Health Issues," *Proceedings of the National Academy of Sciences* 113, no. 42 (2016): 11811–16."

chapter 4 developed a proposal for an adversarial science court, which would scrutinize opposing scientific views in front of a citizen jury. The COVID-19 pandemic provided many examples of the public role of scientific dissent, illustrating the different forms that dissent could take and range of responses that it would receive. I did not find instances of scientific advisory committees offering written dissenting opinions, so I will highlight three other modes of dissent that shared the same basic spirit and aims.

One form of dissent involved individual scientists who set out to challenge the dominant view on COVID-19 risks through their own scientific research. One of the most prominent examples of this category were epidemiologist John Ioannidis and his coauthors, who ran a study that involved giving antibody tests to residents of Santa Clara County, California, and found that COVID-19 infection rates were far higher than believed. This implied that the disease's death rate must be much lower than believed—around the same as influenza. The authors used these findings to criticize strict lockdown policies as an unprecedented evidence fiasco.

A second form of dissent was more organized: three scientists from Harvard, Oxford, and Stanford banded together to produce a declaration that criticized strict lockdown policies, and circulated it for signatures. They claimed that lockdown policies were producing devastating effects, including lower childhood vaccination rates, worsening cardiovascular disease outcomes, fewer cancer screenings, and deteriorating mental health. They proposed a more targeted strategy of isolating and protecting the vulnerable while allowing others to build up immunity through natural infection. The so-called Great Barrington Declaration was signed by over fifty thousand medical and public health scientists and practitioners, and was presented on a website alongside explainer videos and further information.

The final form of dissent that I want to highlight comes closest to my proposal for formal dissents on scientific advisory committees and thus is my personal favorite. A group of prominent UK scientists established an alternative scientific advisory group in response to the failures of the government's official Scientific Advisory Group for Emergencies (SAGE). Members of the rival group, called the Independent SAGE, did not single out a specific scientific claim or policy position to challenge but instead aimed more generally to counteract the official SAGE's lack of accountability and the government's mishandling of the pandemic response. Its main emphasis was on the importance of putting scientific advice in the public domain to ensure that citizens could engage with the reasoning behind the government's strategy. Some of its meetings were livestreamed on YouTube, and all its advice was shared openly with the government and public.

Each of these dissenters took a clear public stance against the dominant scientific advice behind government policies. In their own ways, each accomplished what dissent is meant to do: they forced other scientists to engage with their challenges, and either rethink their positions or—more often—defend them more vigorously and with better evidence. These exchanges were valuable for democratic debate. The antilockdown views initially did not have vocal defenders who also engaged with the scientific models behind government policies. The one country that rejected lockdowns—Sweden—was relentlessly criticized by scientists. Both Ioannidis's study and the Great Barrington Declaration brought attention to the question of the sufficiency of evidence for lockdown policies along with the value trade-offs that were being made to justify them. In doing so, they offered better representation for a significant political perspective that was held by some citizens but frequently dismissed as antiscience. An additional democratic contribution of the Independent SAGE was that it supplied valuable scientific analyses for the opposition parties, which the latter used to criticize the government's response and suggest alternatives.

Despite these contributions, the first two groups of dissenters used some problematic strategies, and their failings are instructive for thinking about better and worse forms of dissent. Ioannidis's study was blasted by the scientific community, not only because it was sloppily done and probably wrong, but because the authors had shared it with the media before going through peer review. Ioannidis's scientific standing gave the study disproportionate attention in the media, which demanded immediate engagement and responses from other scientists. Scientists naturally resented the pressure to engage through the media with a study that had avoided the normal processes of scientific scrutiny and quality control. It was particularly unfair for a well-established scientist to short-circuit these processes at a time when the scientific community faced extraordinary pressure to speed them up, given the high stakes and urgent need for new scientific knowledge.

The Great Barrington Declaration was problematic in a different way. The authors' message straddled scientific claims about the health impacts of lockdowns versus alternative policy approaches, and value judgments about the preferability of different distributions of harms and benefits across groups. These assertions, however, were made without offering scientific evidence. If the document was meant to contest the scientific assumptions behind the government's policies, then its lack of scientific evidence was problematic, and its reliance on signatories was out of place. But it was clear that the document could not have been intended merely as a moral and political statement since the success of its argument hinged on the largely scientific claim that an alternative approach would improve health outcomes. While Ioannidis violated the internal norms of the scientific community, the Great Barrington Declaration blurred the line between the norms of science and those of advocacy, creating the impression that a large number of signatures was marshaled as evidence for the correctness of a scientific view.

Dissenters made important contributions to the scrutiny of the science behind government pandemic policies, but they also pressed some inadequately supported scientific claims and undisclosed value judgments. Dissent is no less valuable for being wrong, yet media hype and a large number of signatures can detract attention from the strengths and weaknesses of the arguments. These issues could have been productively examined in the more formal setting of a science court, following the rules proposed in chapter 4. In fact, given the disagreement between experts, ongoing uncertainty about the science, and value judgments on both sides, the debate between supporters and opponents of lockdown policies would have been ideal material for the court.

Funding and Restricting Science

On December 30, 2019, Shi Zhengli, a Chinese virologist known as "bat woman" for her virus-hunting expeditions in bat caves, received a phone call from the Wuhan Center for Disease Control and Prevention. The center had detected a novel coronavirus in two hospital patients and wanted Shi's lab, the Wuhan Institute of Virology, to investigate. While Shi's research had prepared her to expect a call of this sort at some point, she said that she never thought it would come from Wuhan, over a thousand miles away from the Yunnan bat caves. She remembered thinking, "Could they have come from our lab?"[16]

Shi's lab was one of a small number in the world conducting dangerous gain-of-function experiments, whose aim is to create highly virulent and transmissible viruses.[17] The experiments involve taking live viruses from animals,

---

16. Jane Qiu, "How China's 'Bat Woman' Hunted Down Viruses from SARS to the New Coronavirus," *Scientific American*, June 1, 2020.

17. David Cyranoski, "Inside the Chinese Lab Poised to Study World's Most Dangerous Pathogens," *Nature News*, February 22, 2017.

and manipulating them in the lab to produce more lethal and contagious versions that carry pandemic potential in humans. The purported aim of this research agenda is to understand these viral transformations in order to be better prepared to prevent and control pandemics. The problem is that the research itself creates a pandemic risk. Accidents occur regularly in labs that study contagious viruses. In the past two decades, US labs had accidents involving the release of smallpox, avian flu, and anthrax viruses, while lab accidents in Singapore, Taipei, and Beijing led to SARS infections in humans.[18] Moreover, US inspectors had investigated Shi's lab in 2018 and warned that it suffered from a shortage of adequately trained personnel to operate it safely.[19]

Gain-of-function research is a perfect example of the high-risk, high-uncertainty research areas that I discussed in chapter 6 whose potentially catastrophic risks raise questions about the wisdom of funding and pursuing them. Researchers in the area claim that their work is crucial for pandemic prevention, while critics maintain that its contribution to prevention is more modest than alleged and most benefits could be attained through less risky experiments.[20] In 2014, hundreds of scientists signed a letter asking for an end to experiments involving pathogens carrying pandemic potential.[21] The Obama administration placed a moratorium on this research, but lifted it three years later without justifying its reasoning.[22]

The current practice is to determine funding based purely on the merits of individual proposals, as with most research.[23] Since gain-of-function experiments do not involve human subjects and IRB rules do not require considering risks beyond the lab, there are no grounds for imposing restrictions during

18. Denise Grady, "Pathogen Mishaps Rise as Regulators Stay Clear," *New York Times*, July 19, 2014; David L. Heymann, R. Bruce Aylward, and Christopher Wolff, "Dangerous Pathogens in the Laboratory: From Smallpox to Today's SARS Setbacks and Tomorrow's Polio-Free World," *Lancet* 363, no. 9421 (2004): 1566–68.

19. "Read the State Department Cable That Launched Claims That Coronavirus Escaped from Chinese Lab," *Washington Post*, July 17, 2020."

20. Marc Lipsitch, "Why Do Exceptionally Dangerous Gain-of-Function Experiments in Influenza?," in *Influenza Virus*, ed. Yohei Yamauchi (New York: Humana Press, 2018), 589–608.

21. "Cambridge Working Group Consensus Statement on the Creation of Potential Pandemic Pathogens (PPPs)," July 14, 2014, http://www.cambridgeworkinggroup.org/.

22. Donald G. McNeil Jr., "White House to Cut Funding for Risky Biological Study," *New York Times*, October 17, 2014; Denise Grady, "Studies of Deadly Flu Virus, Once Banned, Are Set to Resume," *New York Times*, March 1, 2019.

23. Lipsitch, "Why Do Exceptionally Dangerous Gain-of-Function Experiments in Influenza?"

ethical review. Biosafety requirements are identified only after the approval of funding, which means that the risks posed by the experiment are not considered in decisions about whether the research should go forward. I argued in chapter 6 that cordoning off the broader impacts of research is morally problematic since scientists are at least partly responsible for foreseeable accidents and possible misuses of their findings on the grounds of negligence or recklessness. I also proposed that funding decisions for such high-risk research must be made more democratically through processes that account for public attitudes toward the risks and benefits alongside expert risk-benefit assessments. The fact that few nonscientists seemed to be aware of the nature and risks of gain-of-function experiments when the COVID-19 pandemic broke out shows that the earlier decision to place this research under a moratorium—and the decision to reverse it later on—was not particularly transparent.

The normative framework that I developed can thus be used to criticize existing procedures and guide future decision-making about funding or restricting research. These theoretical arguments are pitched at a certain level of abstraction, though. The complex political dynamics that came into play over the COVID-19 lab escape hypothesis showed the challenges of applying theoretical arguments to a particular context. Specifically, these dynamics revealed how difficult it can be to acquire reliable answers to the key scientific questions required for decision-making in a context where most actors claiming to speak for the public interest have competing interests and aims, hence casting doubt on their credibility. This fascinating episode is worth discussing in some detail.

For several months in the early stages of the pandemic, there was a controversy over the possibility that the virus may have escaped from the Wuhan Virology Institute. If the theory were true, it would have devastating implications for the assignment of blame and the future of research in the area. The global public had a stake in knowing the origins of the pandemic as well as in decisions about the future of pandemic-potential research. These were fundamentally political questions both because Shi's research was publicly funded—by the US and Chinese governments—and because its risks and impacts fell on literally everyone in the world.

Still, perhaps the greatest damage to the possibility of a sound discussion over the origins of the pandemic and funding of gain-of-function research was the Trump administration's seizing on the lab escape hypothesis without any evidence. In an unusual move, the National Institutes of Health also acted to withdraw a grant that it had already awarded to a nonprofit called the EcoHealth Alliance because it had been collaborating with the Wuhan Virology Institute and indirectly funding its research for many years. A precautionary move that may have been defensible with the appropriate intention and justification became immediately suspect given that it seemed motivated by the Trump administration's desire to deflect blame for its own mishandling of the pandemic response by pointing to China and playing up anti-Chinese sentiment.

In response to the withdrawal of funding from the EcoHealth Alliance, seventy-seven Nobel Laureates signed a letter condemning political interference with the conduct of science and contended that this move would shake public trust in the process of awarding federal funds in research.[24] They went on to claim that it was essential to fund this research in order to control the pandemic and prevent subsequent ones. This letter was problematic in several ways too. It failed to acknowledge the legitimate stake of the administration in the safety of virus research and justifiability of continued support for this research agenda. Nor did it address any concerns about the safety risks of gain-of-function research or the possibility that the Wuhan lab may be implicated in the origins of the pandemic. It simply asserted on authority that this particular grant was vital for pandemic prevention, without explanation or justification. Another letter signed by thirty-one scientific societies demanded more transparency about the decision and accused the administration of politicizing science, thus failing to acknowledge the intrinsically political nature of the conflict.

Against the background of the Trump administration's hostility toward science, these letters may have been strategically defensible. Nevertheless, this political showdown between scientists and the administration was not helpful for those who wanted information on the likelihood of the lab escape hypothesis, the relationship between the EcoHealth Alliance and Shi's lab, and public health implications of funding or failing to fund this research. Few scientists directly engaged with these questions in public, and most mainstream media organizations followed prominent virologists in treating the lab escape hypothesis as a mere rumor, speculation, or conspiracy, without discussing the available evidence. There were also misleading references to a consensus among scientists that the virus was of natural origins. Since only a few scientific articles had been published on the subject, the reference to a consensus

24. "Nobel Laureates and Science Groups Demand NIH Review Decision to Kill Coronavirus Grant," *Science News*, May 21, 2020.

could only mean that many scientists were convinced by the evidence that these few articles offered. This is quite different than a scientific consensus in which many different scientific studies independently confirm the same conclusion.

From an ordinary citizen's perspective, finding reliable answers to the relevant questions was difficult. What was the evidence that the lab escape theory was false? How certain was it? Was it a good idea for the US government to continue to fund gain-of-function research, including collaborations with the Wuhan Virology Institute? As I argued in chapters 5 and 6, these types of questions require political scrutiny and input rather than just scientific competence. The Trump administration, however, did not take the lead in pursuing these questions in good faith, and scientists working in the area could not be fully trusted either. A vigorous public defense of the Wuhan Virology Institute from its longtime collaborator Peter Daszak of the EcoHealth Alliance was hardly credible, even if sincere. Scientists working in the area were not well positioned to make the case since they stood to gain thousands of dollars of research money by defending the importance and safety of this type of research.

This is a good illustration of the funding problem that I took up in more abstract terms in chapter 5. A Kuhnian model of science implies that scientists working under a particular paradigm—or a research program such as gain-of-function research—will be committed to its expansion through continued funding and will actively resist threats to it. As a result, giving autonomy to specialists over the distribution of large amounts of funding can make it difficult for alternative approaches to receive funding and support, which in turn makes it difficult to challenge the dominant paradigm or withdraw its funding. While scientists working in a paradigm will understand the merits of the science best, they will not be most reliable for assessing its social value and cannot be entrusted with the task of weighing its value against the risks.

Throughout the book, I emphasized the important role that publicly oriented scientific dissent can play in facilitating decision-making with and about science, especially under conditions of uncertainty and disagreement. The controversy over the origins of COVID-19 provided further illustrations of the benefits of dissent in exposing the limits of existing views, and offering new evidence and arguments. This episode also supplied evidence of an additional service that dissent can provide for democratic debate—one that I have not highlighted so far. It showed how the existence of serious and credible dissenting views can elevate the credibility of the whole debate in the eyes of a public audience, including the credibility of other actors whose conflicts of interest would have made them less credible otherwise.

To illustrate the role of dissent in this case, it is helpful to first say something about the nature of the available evidence six months after the outbreak. The main evidence against the lab hypothesis at this point consisted in the fact that the virus had evolved certain unusual features in its receptor-binding domain and spike proteins that were optimized to infect humans.[25] While the possibility that these mutations happened in lab cell cultures could not be ruled out, the Wuhan lab had not previously published the discovery of a strain that was genetically similar enough to SARS-CoV-2 to make passaging in cell cultures plausible in a reasonable time frame. For this hypothesis to be true, they would have had to discover a highly similar strain, and conduct passaging experiments without publishing or sharing any of their findings. Natural selection in an animal host was the most probable and parsimonious explanation of these mutations given the degree of expected though as yet undiscovered diversity in nature. All this together formed a persuasive case against this theory, but hardly a conclusive one. The discovery of an animal in the wild carrying an almost identical virus strain would be more decisive evidence, but no such animal had been found.[26]

While most scientists found the existing level of evidence sufficient to rule out a lab escape, Alina Chan, a young postdoctoral researcher at the Broad Institute at MIT, began to pursue this hypothesis because she thought the virus's almost perfect adaptation to the human body presented a striking contrast with earlier SARS viruses, which had mutated rapidly after entering human hosts in order to adapt to their environment. She argued that this made it plausible that the virus could have been passaged in human cells in a lab.[27] Other circumstantial evidence, such as the Chinese government's refusal to allow any investigations of the lab and the fact that the virus's nearest-known genetic relative had been sequenced in that lab several years ago but published only after the pandemic, lent some credence to this hypothesis too. While many conspiracy theories on the subject were circulating at the time, Chan's

25. Kristian G. Andersen, Andrew Rambaut, W. Ian Lipkin, Edward C. Holmes, and Robert F. Garry, "The Proximal Origin of SARS-CoV-2," *Nature Medicine* 26 (2020): 450–52.

26. David Cyranoski, "The Biggest Mystery: What It Will Take to Trace the Coronavirus Source," *Nature News*, June 5, 2020.

27. Rowan Jacobsen, "Could COVID-19 Have Escaped from a Lab?," *Boston Magazine*, September 9, 2020.

pursuit was set apart by her scientific credentials and the fact that she seemed to have no stake in the matter except for her interest in it as a scientific puzzle. She was definitely not a Trump supporter and did not have a record of opposing gain-of-function research, unlike some other vocal scientists.

For those who followed her work, her pursuit brought much clarity to the state of the debate, pointing out the limits of what could be inferred from the existing evidence. To put it in Millian terms, her work first exposed the limits of the evidence for the majority view as well as the evidence that would be needed to settle the debate. It showed that a lab escape was a live possibility, though a highly unlikely one. Second, it moved the conversation along by supplying some interesting new observations based on comparisons with earlier SARS epidemics, which defenders of the natural origins hypothesis had to engage with. Third, it allowed members of the public to understand the grounds of the majority perspective better, through careful engagement with studies rather than authoritative assertion and dismissal. This allowed for a more nuanced position against the Trump administration's baseless claims, while leaving open the possibility that this line of research may well have been too risky and the Chinese government's secrecy over the research at the lab justified mistrust. Finally, its perhaps not undesirable result was that it ended up making the majority view seem more convincing and credible, as it had been truly tested against the best reconstruction of the alternative versus asserted on authority. For those who were simply concerned about the possibility of a lab accident, it was reassuring to know that a scientist well qualified to examine the evidence carefully pursued the possibility.

There is always a risk that a dissenting view that happens to align with a position pushed by conspiracy theories will fuel public misinformation. The opinion that the scientific and democratic benefits of dissent are not worth this risk is unfortunately becoming all too common. It is true that Chan's work gave a careful and scientific defense of a position whose most visible defenders were conspiracists and the Trump administration. At the same time, arguing that she should not have pursued this line of thought assumes that we know for sure that she was mistaken. I tried to show that this was not the case; there was not enough scientific evidence on the origins of the COVID-19 pandemic to rule out the lab escape hypothesis. Under genuine uncertainty, I don't think the fact that there are many ill-motivated defenders of a scientific claim is enough reason to abandon well-motivated defenses of it, especially if it would make an enormous difference if the claim turned out to be true.

Conspiracists feed off distrust of mainstream institutions, whereas their critics believe that these institutions merit trust.[28] In the controversy over the origins of the COVID-19 pandemic, it was difficult to say whether distrust was out of place. The Chinese government repeatedly refused to allow investigations into the Wuhan Virology Institute and silenced scientists in the early days of the pandemic. The line between justified and conspiratorial mistrust becomes even more blurred when an authoritarian government is in the fray. In such a case, dissenting scientists can render an important public service by investigating the concerns of those who mistrust authorities. And unlike conspiracists, they can do so by relying on the available evidence instead of unsubstantiated or unfalsifiable assertions. The evidence and arguments of a trustworthy dissenter could persuade citizens who share conspiracists' distrust of authorities, but not their immunity to refutation. If this is true, then discouraging dissenting views in public in order to fight conspiracy theories would be doubly wrong: it would prevent the expression of a healthy scientific skepticism that could contribute to the discovery of truth and it would block the possibility of earning citizens' trust by showing that their suspicions are examined in good faith by someone competent to do so rather than dismissed on authority.

28. Nancy L. Rosenblum and Russell Muirhead, *A Lot of People Are Saying: The New Conspiracism and the Assault on Democracy* (Princeton, NJ: Princeton University Press, 2019).

REFERENCES

Abizadeh, Arash. "On the Demos and Its Kin: Nationalism, Democracy, and the Boundary Problem." *American Political Science Review* 106, no. 4 (2012): 867–82.

Abramson, Jeffrey. *We, the Jury: The Jury System and the Ideal of Democracy.* New York: Basic Books, 1994.

Amar, Akhil. "The Central Meaning of Republican Government: Popular Sovereignty, Majority Rule, and the Denominator Problem." *University of Colorado Law Review* 65 (1993): 749–87.

———. "Reinventing Juries: Ten Suggested Reforms." *UC Davis Law Review* 28 (1995): 1169–95.

Andersen, Kristian G., Andrew Rambaut, W. Ian Lipkin, Edward C. Holmes, and Robert F. Garry. "The Proximal Origin of SARS-CoV-2." *Nature Medicine* 26 (2020): 450–52.

Anderson, Elizabeth. "Feminist Epistemology: An Interpretation and a Defense." *Hypatia* 10, no. 3 (1995): 50–84.

———. "Knowledge, Human Interests, and Objectivity in Feminist Epistemology." *Philosophical Topics* 23, no. 2 (Fall 1995): 27–58.

———. "The Epistemology of Democracy." *Episteme* 3, nos. 1–2 (2006): 8–22.

Anderson, Kenneth, and Matthew C. Waxman. "Debating Autonomous Weapon Systems, Their Ethics, and Their Regulation under International Law." In *The Oxford Handbook of Law, Regulation and Technology*, edited by Roger Brownsword, Eloise Scotford, and Karen Yeung, 1097–118. Oxford: Oxford University Press, 2016.

Angell, Marcia. *Science on Trial.* New York: W. W. Norton and Company, 1996.

Arkin, Ronald. "Lethal Autonomous Systems and the Plight of the Noncombatant." *AISB Quarterly* 137 (2013).

"Autonomous Weapons: An Open Letter from AI and Robotics Researchers." https://futureoflife.org/open-letter-autonomous-weapons.

Avery, Christopher, William Bossert, Adam Clark, Glenn Ellison, and Sara Fisher. "An Economist's Guide to Epidemiology Models of Infectious Disease." *Journal of Economic Perspectives* 34, no. 4 (Fall 2020): 79–104.

Avin, Shahar. "Centralized Funding and Epistemic Exploration." *British Journal for the Philosophy of Science* 70, no. 3 (2017): 629–56.

———. "Mavericks and Lotteries." *Studies in History and Philosophy of Science Part A* 76 (2019): 13–23.

Bachrach, Peter, and Morton Baratz. "Two Faces of Power." *American Political Science Review* 56, no. 4 (1962) 947–52.

Bächtiger, André, Maija Setälä, and Kimmo Grönlund, "Towards a New Era of Deliberative Mini-Publics." In *Deliberative Mini-Publics: Innovating Citizens in the Democratic Process,* edited by Kimmo Grönlund, André Bächtiger, and Maija Setälä, 225–47. Colchester, UK: ECPR Press, 2014.

Bagg, Samuel. "Can Deliberation Neutralise Power?" *European Journal of Political Theory* 17, no. 3 (2018): 257–79.

———. "The Power of the Multitude: Answering Epistemic Challenges to Democracy," *American Political Science Review* 112, no. 4 (November 2018): 891–904.

Barry, Brian. "Is Democracy Special?" In *Democracy and Power: Essays in Political Theory, Part 1,* 24–60. Oxford: Clarendon Press, 1991.

———. *Justice as Impartiality: A Treatise on Social Justice, Vol 2.* Oxford: Oxford University Press, 1995.

Beck, Silke. "Between Tribalism and Trust: The IPCC under the 'Public Microscope.'" *Nature and Culture* 7, no. 2 (2012): 151–73.

Beck, Ulrich. *Risk Society: Towards a New Modernity.* London: Sage, 1992.

Beck, Ulrich, and Peter Wehling. "The Politics of Non-Knowing: An Emerging Area of Social and Political Conflict in Reflexive Modernity." In *The Politics of Knowledge,* edited by Fernando Domínguez Rubio and Patrick Baert, 33–57. London: Routledge, 2012.

Benhabib, Seyla. *The Rights of Others: Aliens, Residents, and Citizens.* Cambridge: Cambridge University Press, 2004.

Bentham, Jeremy. *Securities against Misrule and Other Constitutional Writings for Tripoli and Greece.* Edited by Philip Schofield. Oxford: Clarendon Press, 1990.

Berg, Paul, David Baltimore, Sydney Brenner, Richard O. Roblin, and Maxine F. Singer. "Summary Statement of the Asilomar Conference on Recombinant DNA Molecules." *Proceedings of the National Academy of Sciences of the United States of America* 72, no. 6 (June 1975): 1981–84.

Berlin, Isaiah. "Two Concepts of Liberty." In *Liberty,* edited by Henry Hardy, 166–217. Oxford: Oxford University Press, 2002.

Betz, Gregor. "In Defence of the Value Free Ideal." *European Journal for Philosophy of Science* 3 (2013): 207–20.

Boffey, Philip M. "Anatomy of a Decision: How the Nation Declared War on Swine Flu." *Science* 192, no. 4240 (May 1976): 636–41.

———. "Experiment Planned to Test Feasibility of a 'Science Court.'" *Science* 193, no. 4248 (July 1976): 129.

Bogner, Alexander. "The Paradox of Participation Experiments." *Science, Technology, and Human Values* 37, no. 5 (2012): 506–27.

Bohman, James. "Democracy as Inquiry, Inquiry as Democratic: Pragmatism, Social Science, and the Cognitive Division of Labor." *American Journal of Political Science* 43, no. 2 (1999): 590–607.

———. "Deliberative Democracy and the Epistemic Benefits of Diversity." *Episteme* 3, no. 3 (2006): 175–91.

Bollinger, Lee C. "The Rationale of Public Regulation of the Media." In *Democracy and the Mass Media,* edited by Judith Lichtenberg, 355–68. Cambridge: Cambridge University Press, 1990.

Bostrom, Nick. "Existential Risks: Analyzing Human Extinction Scenarios and Related Hazards." *Journal of Evolution and Technology* 9, no. 1 (2002).

Bratman, Michael E. "Practical Reasoning and Acceptance in a Context." *Mind* 101, no. 401 (1992): 1–15.

Braun, Kathrin, and Susanne Schultz. "'. . . a Certain Amount of Engineering Involved': Constructing the Public in Participatory Governance Arrangements." *Public Understanding of Science* 19, no. 4 (2010): 403–19.

Brennan, Jason. *Against Democracy.* Princeton, NJ: Princeton University Press, 2016.

Brennan, Patricia. "Why I Study Duck Genitalia." *Slate*, April 2, 2013. http://www.slate.com /articles/health_and_science/science/2013/04/duck_penis_controversy_nsf_is_right _to_fund_basic_research_that_conservatives.html.

Brennan, William J. "In Defense of Dissents." *Hastings Law Journal* 37 (1985): 427–39.

Brighouse, Harry. "Neutrality, Publicity, and State Funding of the Arts." *Philosophy and Public Affairs* 24, no. 1 (1995): 35–63.

Bright, Liam K. "Du Bois' Democratic Defence of the Value Free Ideal." *Synthese* 195, no. 5 (2018): 2227–45.

Broad, William J. "Billionaires with Big Ideas Are Privatizing Science." *New York Times*, March 15, 2014. http://www.nytimes.com/2014/03/16/science/billionaires-with-big-ideas-are -privatizing-american-science.html?_r=0.

Brown, Mark B. *Science in Democracy: Expertise, Institutions, and Representation.* Cambridge, MA: MIT Press, 2009.

———. "Philip Kitcher, *Science in a Democratic Society*," *Minerva* 51 (2013): 389–97.

———. "Expertise and Deliberative Democracy." In *Deliberative Democracy: Issues and Cases*, edited by Stephen Elstub and Peter McLaverty, 50–69. Edinburgh: Edinburgh University Press, 2014.

Brown, Matthew J., and Joyce C. Havstad. "The Disconnect Problem, Scientific Authority, and Climate Policy." *Perspectives on Science* 25, no. 1 (2017): 67–94.

Bruno, Jonathan. *"Democracy beyond Disclosure: Secrecy, Transparency, and the Logic of Self-Government."* PhD diss., Harvard University, 2017.

———. "Vigilance and Confidence: Jeremy Bentham, Publicity, and the Dialectic of Political Trust and Distrust." *American Political Science Review* 111, no. 2 (2017): 295–307.

Buck, Holly Jean. "Climate Engineering: Spectacle, Tragedy or Solution? A Content Analysis of News Media Framing." In *Interpretative Approaches to Global Climate Governance: (De)constructing the Greenhouse*, edited by Chris Methmann, Delf Rothe, and Benjamin Stephan, 166–83. New York: Routledge, 2013.

Buck, Holly Jean, Andrea R. Gammon, and Christopher J. Preston. "Gender and Geoengineering." *Hypatia* 29, no. 3 (2014): 651–69.

Burns, Wil, and Andrew Strauss, eds. *Climate Change Geoengineering: Philosophical Perspectives, Legal Issues, and Governance Frameworks.* Cambridge: Cambridge University Press, 2013.

Bush, Vannevar. *Science: The Endless Frontier.* Washington, DC: National Science Foundation, 1960. First published 1945.

Cain, Bruce E. "Redistricting Commissions: A Better Political Buffer?" *Yale Law Journal* 121 (2012): 1808–44.

Callon, Michel, Pierre Lascoumes, and Yannick Barthe. *Acting in an Uncertain World: An Essay on Technical Democracy.* Cambridge, MA: MIT Press, 2009.

"Cambridge Working Group Consensus Statement on the Creation of Potential Pandemic Pathogens (PPPs)." July 14, 2014. http://www.cambridgeworkinggroup.org/.

Carroll, Noël. "Can Government Funding of the Arts Be Justified Theoretically?" *Journal of Aesthetic Education* 21, no. 1 (Spring 1987): 21–35.

Cartwright, Nancy. *How the Laws of Physics Lie.* New York: Oxford University Press, 1983.

———. *The Dappled World: A Study of the Boundaries of Science.* Cambridge: Cambridge University Press, 1999.

Cartwright, Nancy, and Jeremy Hardie. *Evidence-Based Policy: A Practical Guide to Doing It Better.* New York: Oxford University Press, 2012.

Casper, Barry, and Paul Wellstone. "The Science Court on Trial in Minnesota." *Hastings Center Report* 8, no. 4 (1978): 5–7.

Chambers, Simone. "Behind Closed Doors: Publicity, Secrecy, and the Quality of Deliberation." *Journal of Political Philosophy* 12, no. 4 (2004): 389–410.

———. "Rhetoric and the Public Sphere: Has Deliberative Democracy Abandoned Mass Democracy?" *Political Theory* 37, no. 3 (2009): 323–50.

———. "Balancing Epistemic Quality and Equal Participation in a System Approach to Deliberative Democracy." *Social Epistemology* 31, no. 3 (2017): 266–76.

Christiano, Thomas. *The Rule of the Many: Fundamental Issues in Democratic Theory.* Boulder, CO: Westview Press, 1996.

———. "Rational Deliberation among Experts and Citizens." In *Deliberative Systems: Deliberative Democracy at the Large Scale,* edited by John Parkinson and Jane Mansbridge, 27–52. Cambridge: Cambridge University Press, 2012.

Churchman, C. West. "Statistics, Pragmatics, Induction." *Philosophy of Science* 15 (1948): 249–68.

———. "Science and Decision Making." *Philosophy of Science* 23, no. 3 (July 1956): 247–49.

Coen, Deborah. *The Earthquake Observers: Disaster Science from Lisbon to Richter.* Chicago: University of Chicago Press, 2013.

Cohen, Joshua. "Deliberation and Democratic Legitimacy." In *The Good Polity: Normative Analysis of the State,* edited by Alan Hamlin and Philip Pettit, 67–92. Oxford: Blackwell, 1989.

Collini, Stefan. *What Are Universities For?* London: Penguin Books, 2012.

Collins, Harry, and Robert Evans. *Rethinking Expertise.* Chicago: University of Chicago Press, 2007.

———. *Why Democracies Need Science.* Cambridge, UK: Polity Press, 2017.

Collins, Harry, Martin Weinel, and Robert Evans. "The Politics and Policy of the Third Wave: New Technologies and Society." *Critical Policy Studies* 4, no. 2 (2010): 185–201.

Cordelli, Chiara. *The Privatized State.* Princeton, NJ: Princeton University Press, 2020.

Crosby, Ned. "Citizens' Juries: One Solution for Difficult Environmental Questions." In *Fairness and Competence in Citizen Participation: Evaluating Models for Environmental Discourse,* edited by Ortwin Renn, Thomas Webler, and Peter Wiedemann, 157–74. Dordrecht: Kluwer Academic Press, 1995.

Cyranoski, David. "Inside the Chinese Lab Poised to Study World's Most Dangerous Pathogens." *Nature News,* February 22, 2017.

———. "The Biggest Mystery: What It Will Take to Trace the Coronavirus Source." *Nature News*, June 5, 2020.

Dahl, Robert A. *Controlling Nuclear Weapons: Democracy versus Guardianship*. Syracuse, NY: Syracuse University Press, 1985.

———. *Democracy and Its Critics*. New Haven, CT: Yale University Press, 1989.

Daniels, Stephen, and Joanne Martin. *Civil Juries and the Politics of Reform*. Evanston, IL: Northwestern University Press, 1996.

Davies, Gareth. "The Psychological Costs of Geoengineering: Why It Might Be Hard to Accept Even If It Works." In *Climate Change Geoengineering: Philosophical Perspectives, Legal Issues, and Governance Frameworks*, edited by Wil Burns and Andrew Strauss, 59–78. Cambridge: Cambridge University Press, 2013.

de Melo-Martín, Inmaculada, and Kristen Intemann. "The Risk of Using Inductive Risk to Challenge the Value-Free Ideal." *Philosophy of Science* 83, no. 4 (2016): 500–520.

———. *The Fight against Doubt: How to Bridge the Gap between Scientists and the Public*. New York: Oxford University Press, 2018.

Dewey, John. *The Public and Its Problems*. Chicago: Swallow Press, 1927.

Diamond, Dan. "Suddenly, Public Health Officials Say Social Justice Matters More Than Social Distance." *Politico*, April 6, 2020.

Dollar, John. "The Man Who Predicted an Earthquake." *Guardian*, April 5, 2010. https://www.theguardian.com/world/2010/apr/05/laquila-earthquake-prediction-giampaolo-giuliani.

Douglas, Heather. "The Irreducible Complexity of Objectivity." *Synthese* 138, no. 3 (2003): 453–73.

———. "The Moral Responsibilities of Scientists (Tensions between Autonomy and Responsibility)." *American Philosophical Quarterly* 40, no. 1 (January 2003): 59–68.

———. *Science, Policy, and the Value-Free Ideal*. Pittsburgh: University of Pittsburgh Press, 2009.

Douglas, Mary, and Aaron Wildavsky. *Risk and Culture: An Essay on the Selection of Technological and Environmental Dangers*. Berkeley: University of California Press, 1982.

Downey, Leah M. "Delegation in Democracy: A Temporal Analysis." *Journal of Political Philosophy* (2020). doi.org/10.1111/jopp.12234.

Dupré, John. *The Disorder of Things: Metaphysical Foundations of the Disunity of Science*. Cambridge, MA: Harvard University Press, 1995.

———. "Science and Values and Values in Science." *Inquiry* 47, no. 5 (2004): 505–14.

———. "Fact and Value." In *Value-Free Science: Ideals and Illusions?*, edited by Harold Kincaid, John Dupré, and Alison Wylie, 27–42. New York: Oxford University Press, 2007.

Dworkin, Ronald. *A Matter of Principle*. Cambridge, MA: Harvard University Press, 1985.

Eisenberg, Theodore, Paula L. Hannaford-Agor, Valerie P. Hans, Nicole L. Waters, and G. Thomas Munsterman. "Judge-Jury Agreement in Criminal Cases: A Partial Replication of Kalven and Zeisel's *The American Jury*." *Journal of Empirical Legal Studies* 2, no. 1 (2005): 171–207.

Eker, Sibel. "Validity and Usefulness of COVID-19 Models." *Humanities and Social Sciences Communications* 7, no. 54 (2020): 1–5.

Elliott, Kevin C., and Ted Richards, eds. *Exploring Inductive Risk: Case Studies of Values in Science*. New York: Oxford University Press, 2017.

Elliott, Kevin J. "Democracy's Pin Factory: Issue Specialization, the Division of Cognitive Labor, and Epistemic Performance." *American Journal of Political Science* 64, no. 2 (2020): 385–97.

Emanuel, Ezekiel J., David Wendler, and Christine Grady. "What Makes Clinical Research Ethical?" *JAMA* 283, no. 20 (2000): 2701–11.

Epstein, Steven. *Impure Science: AIDS, Activism, and the Politics of Knowledge.* Berkeley: University of California Press, 1996.

———. *Inclusion: The Politics of Difference in Medical Research.* Chicago: University of Chicago Press, 2008.

Estlund, David. *Democratic Authority: A Philosophical Framework.* Princeton, NJ: Princeton University Press, 2008.

ETC Group. "UK Royal Society on Geoengineering: The Emperor's New Climate?" Press release, August 28, 2009.

Evans, John H., and Eszter Hargittai. "Who Doesn't Trust Fauci? The Public's Belief in the Expertise and Shared Values of Scientists in the COVID-19 Pandemic." *Socius* 6 (2020). doi .org/10.1177/2378023120947337.

Ezrahi, Yaron. "Utopian and Pragmatic Rationalism: The Political Context of Scientific Advice." *Minerva* 18, no. 1 (1980): 111–31.

Fang, Ferric C., and Arturo Casadevall. "Research Funding: The Case for a Modified Lottery." *mBio* 7, no. 3 (2016): 1–7.

Farber, Henry, and Michelle White. *Medical Malpractice: An Empirical Examination of the Litigation Process.* Cambridge, MA: National Bureau of Economic Research, 1990.

Federal Rules of Evidence. Grand Rapids, MI: Michigan Legal Publishing Ltd., 2021.

Feinberg, Joel. *The Moral Limits of the Criminal Law Volume 3: Harm to Self.* New York: Oxford University Press, 1989.

———. "Not with My Tax Money: The Problem of Justifying Government Subsidies for the Arts." *Public Affairs Quarterly* 8, no. 2 (1994): 101–23.

Feinstein, Alvin. "An Additional Basic Science for Clinical Medicine: II. The Limitations of Randomized Trials." *Annals of Internal Medicine* 99, no. 4 (1983): 544–50.

Feyerabend, Paul. "Consolations for the Specialist." In *Criticism and the Growth of Knowledge,* edited by Imre Lakatos and Alan Musgrave, 197–203. London: Cambridge University Press, 1970.

Fishkin, James. *Democracy and Deliberation.* New Haven, CT: Yale University Press, 1991.

———. *The Voice of the People: Public Opinion and Democracy.* New Haven, CT: Yale University Press, 1997.

Flegal, Jane, and Aarti Gupta. "Evoking Equity as a Rationale for Solar Geoengineering Research? Scrutinizing Emerging Expert Visions of Equality." *International Environmental Agreements: Politics, Law and Economics* 18, no. 1 (2018): 45–61.

Fleming, James. *The Checkered History of Weather and Climate Control.* New York: Columbia University Press, 2012.

Fuerstein, Michael. "Epistemic Trust and Liberal Justification." *Journal of Political Philosophy* 21, no. 2 (2013): 179–99.

Fuller, Jonathan. "Models v. Evidence." *Boston Review,* May 5, 2020. http://bostonreview.net /science-nature/jonathan-fuller-models-v-evidence.

Fung, Archon. "Survey Article: Recipes for Public Spheres: Eight Institutional Design Choices and Their Consequences." *Journal of Political Philosophy* 11, no. 3 (2003): 338–67.

————. *Empowered Participation: Reinventing Urban Democracy.* Princeton, NJ: Princeton University Press, 2009.

Funtowicz, Silvio, and Jerome Ravetz. "Science for the Post-Normal Age." *Futures* 25, no. 7 (1993): 739–55.

Gardiner, Stephen. "A Core Precautionary Principle." *Journal of Political Philosophy* 14, no. 1 (2006): 33–60.

Gastil, John, and Peter Levine, eds. *The Deliberative Democracy Handbook: Strategies for Effective Civic Engagement in the Twenty-First Century.* San Francisco: Jossey-Bass, 2005.

Gastil, John, Robert C. Richards Jr., and Katherine R. Knobloch. "Vicarious Deliberation: How the Oregon Citizens' Initiative Review Influenced Deliberation in Mass Elections." *International Journal of Communication* 8, no. 1 (2014): 62–89.

Geertz, Clifford. "Thick Description: Toward an Interpretive Theory of Culture." In *Readings in the Philosophy of Social Science,* edited by Michael Martin and Lee McIntyre, 213–32. Cambridge, MA: MIT Press, 1994.

Gieryn, Thomas F. "Boundary-Work and the Demarcation of Science from Non-Science: Strains and Interests in Professional Ideologies of Scientists." *American Sociological Review* 48, no. 6 (December 1983): 781–95.

Gillispie, Charles C. *Science and Polity in France: The Revolutionary and Napoleonic Years.* Princeton, NJ: Princeton University Press, 2004.

Glanz, James, Gaia Pianigiani, Jeremy White, and Karthik Patanjali. "Genoa Bridge Collapse: The Road to Tragedy." *New York Times,* September 6, 2018. https://www.nytimes.com /interactive/2018/09/06/world/europe/genoa-italy-bridge.html.

Goldman, Alvin. "Experts: Which Ones Should You Trust?" *Philosophy and Phenomenological Research* 63, no. 1 (2001): 85–110.

Goodin, Robert. "Excused by the Unwillingness of Others?" *Analysis* 72, no. 1 (2012): 18–24.

Goodin, Robert, and Kai Spiekermann. *An Epistemic Theory of Democracy.* Oxford: Oxford University Press, 2018.

Graber, Frédéric. "Obvious Decisions: Decision-making among French Ponts-et-Chaussées Engineers around 1800." *Social Studies of Science* 37, no. 6 (2007): 935–60.

Grady, Denise. "Pathogen Mishaps Rise as Regulators Stay Clear." *New York Times,* July 19, 2014.

————. "Studies of Deadly Flu Virus, Once Banned, Are Set to Resume." *New York Times,* March 1, 2019.

Graves, Nicholas, Adrian G. Barnett, and Philip Clarke. "Funding Grant Proposals for Scientific Research: Retrospective Analysis of Scores by Members of Grant Review Panel." *BMJ* 343 (2011). https://www.bmj.com/content/343/bmj.d4797.

Greenberg, Daniel S. "Chance and Grants." *Lancet* 351 (1998): 686.

————. *Science for Sale: The Perils, Rewards, and Delusions of Campus Capitalism.* Chicago: University of Chicago Press, 2007.

Greene, Benjamin. *Eisenhower, Science Advice, and the Nuclear Test-Ban Debate, 1945–1963.* Stanford, CA: Stanford University Press, 2007.

Greenfieldboyce, Nell. "'Shrimp on a Treadmill': The Politics of 'Silly' Studies." NPR, August 23, 2011. https://www.npr.org/2011/08/23/139852035/shrimp-on-a-treadmill-the-politics-of -silly-studies.

Grey, Stephen, and Andrew MacAskill. "Special Report: Johnson Listened to His Scientists about Coronavirus—but They Were Slow to Sound the Alarm." Reuters, April 7, 2020.

Guerrero, Alexander. "Living with Ignorance in a World of Experts." In *Perspectives on Ignorance from Moral and Social Philosophy*, edited by Rik Peels, 156–85. New York: Routledge, 2017.

Guinier, Lani. "Demosprudence through Dissent." *Harvard Law Review* 122 (2008): 6–137.

Guston, David. *Between Politics and Science: Assuring the Integrity and Productivity of Research.* Cambridge: Cambridge University Press, 2000.

———. "On Consensus and Voting in Science: From Asilomar to the National Toxicology Program." In *The New Political Sociology of Science: Institutions, Networks, and Power*, edited by Scott Frickel and Kelly Moore, 378–404. Madison: University of Wisconsin Press, 2006.

Gutmann, Amy, and Dennis Thompson. *Democracy and Disagreement.* Cambridge, MA: Harvard University Press, 1996.

———. *Why Deliberative Democracy?* Princeton, NJ: Princeton University Press, 2004.

Habermas, Jürgen. *Knowledge and Human Interests.* Translated by Jeremy Shapiro. Cambridge, UK: Polity Press, 1987.

———. *Toward a Rational Society: Student Protest, Science, and Politics.* Translated by Jeremy Shapiro. Cambridge, UK: Polity Press, 1987.

———. *Between Facts and Norms.* Translated by William Rehg. Cambridge, MA: MIT Press, 1996.

———. *On the Pragmatics of Social Interaction.* Translated by Barbara Fultner. Cambridge, MA: MIT Press, 2001.

Hall, Stephen S. "Scientists on Trial: At Fault?" *Nature*, September 14, 2011. https://www.nature.com/news/2011/110914/full/477264a.html.

Hamburger, Philip. "IRB Licensing." In *Who's Afraid of Academic Freedom?*, edited by Akeel Bilgrami and Jonathan Cole, 153–90. New York: Columbia University Press, 2015.

Hannaford, Paula L., Valerie P. Hans, and G. Thomas Munsterman. "Permitting Jury Discussions during Trial: Impact of the Arizona Reform." *Law and Human Behavior* 24 (2000): 359–82.

Hans, Valerie. "Jury Systems around the World." *Annual Review of Law and Social Science* 4 (2008): 275–97.

Hardin, Russell. "Street-Level Epistemology and Democratic Participation." *Journal of Political Philosophy* 10, no. 2 (2002): 212–29.

Hauray, Boris, and Philippe Urfalino. "Mutual Transformation and the Development of European Policy Spaces: The Case of Medicines Licensing." *Journal of European Public Policy* 16, no. 3 (2009): 431–49.

Havstad, Joyce C., and Matthew J. Brown. "Inductive Risk, Deferred Decisions, and Climate Science Advising." In *Exploring Inductive Risk: Case Studies of Values in Science*, edited by Kevin Elliott and Ted Richards, 101–25. New York: Oxford University Press, 2017.

Hayek, Friedrich August von. "The Use of Knowledge in Society." *American Economic Review* 35, no. 4 (1945): 519–30.

Hersh, Eitan D., and Matthew N. Goldenberg. "Democratic and Republican Physicians Provide Different Care on Politicized Health Issues." *Proceedings of the National Academy of Sciences* 113, no. 42 (2016): 11811–16.

Hess, David J. *Undone Science: Social Movements, Mobilized Publics, and Industrial Transitions.* Cambridge, MA: MIT Press, 2016.

Heuer, Larry, and Steven Penrod. "Trial Complexity: A Field of Investigation of Its Meaning and Effects." *Law and Human Behavior* 18 (1994): 29–51.

Heymann, David L., R. Bruce Aylward, and Christopher Wolff. "Dangerous Pathogens in the Laboratory: From Smallpox to Today's SARS Setbacks and Tomorrow's Polio-Free World." *Lancet* 363, no. 9421 (2004): 1566–68.

Heyward, Clare. "Is There Anything New under the Sun? Exceptionalism and Novelty in Debating Geoengineering Governance." In *The Ethics of Climate Change Governance,* edited by Aaron Maltais and Catriona McKinnon, 135–54. London: Rowman and Littlefield, 2015.

Holst, Cathrine, and Anders Molander. "Epistemic Democracy and the Role of Experts." *Contemporary Political Theory* 18, no. 4 (2019): 541–61.

Holst, Cathrine, and Silje H. Tørnblad. "Variables and Challenges in Assessing EU Experts' Performance." *Politics and Governance* 3, no. 1 (2015): 166–78.

Hudson, Jennifer M. "The Bureaucratic Mentality in Democratic Theory and Contemporary Democracy." PhD diss., Columbia University, 2016.

Hulme, Mike. *Can Science Fix Climate Change? A Case against Climate Engineering.* Cambridge, UK: Polity Press, 2014.

Ingham, Sean. *Rule by Multiple Majorities: A New Theory of Popular Control.* Cambridge: Cambridge University Press, 2019.

Intergovernmental Panel on Climate Change. "Organization." https://archive.ipcc.ch /organization/organization.shtml.

Ioannidis, John P. A. "The Totality of the Evidence." *Boston Review,* May 26, 2020. http:// bostonreview.net/science-nature/john-p-ioannidis-totality-evidence.

Ioannidis, John P. A., Sally Cripps, and Martin A. Tanner. "Forecasting for COVID-19 Has Failed." *International Journal of Forecasting* (forthcoming).

Irwin, Alan, and Brian Wynne. *Misunderstanding Science? The Public Reconstruction of Science and Technology.* Cambridge: Cambridge University Press, 1996.

Irzık, Gürol, and Faik Kurtulmuş. "Well-Ordered Science and Public Trust in Science." *Synthese* (2018). doi.org/10.1007/s11229-018-02022-7.

Ivkovic, Sanja K., and Valerie P. Hans. "Jurors' Evaluations of Expert Testimony: Judging the Messenger and the Message." *Law and Social Inquiry* 28 (2003): 441–82.

Jacobsen, Rowan. "Could COVID-19 Have Escaped from a Lab?" *Boston Magazine,* September 9, 2020.

Jasanoff, Sheila. *The Fifth Branch: Science Advisers as Policymakers.* Cambridge, MA: Harvard University Press, 1990.

———, ed. *States of Knowledge: The Co-Production of Science and Social Order.* London: Routledge, 2004.

Jasanoff, Sheila, and Sang-Hyun Kim. "Containing the Atom: Sociotechnical Imaginaries and Nuclear Power in the United States and South Korea." *Minerva* 47, no. 2 (2009): 119–46.

———, eds. *Dreamscapes of Modernity: Sociotechnical Imaginaries and the Fabrication of Power.* Chicago: University of Chicago Press, 2015.

Jewett, Andrew. *Science under Fire: Challenges to Scientific Authority in Modern America.* Cambridge, MA: Harvard University Press, 2020.

John, Stephen. "Inductive Risk and the Contexts of Communication." *Synthese* 192, no. 1 (2015): 79–96.

———. "Epistemic Trust and the Ethics of Science Communication: Against Transparency, Openness, Sincerity and Honesty." *Social Epistemology* 32, no. 2 (2018): 75–87.

Joss, Simon, and John Durant, eds. *Public Participation in Science: The Role of Consensus Conferences in Europe.* London: Science Museum, 1995.

Jurs, Andrew. "Science Court: Past Proposals, Current Considerations, and a Suggested Structure." *Virginia Journal of Law and Technology* 15, no. 1 (2010): 1–43.

Kahan, Dan M., Paul Slovic, Donald Braman, and John Gastil. "Fear of Democracy: A Cultural Evaluation of Sunstein on Risk." *Harvard Law Review* 119 (2006): 1071–110.

Kalven, Harry and Hans Zeisel. *The American Jury.* Boston: Little, Brown and Company, 1966.

Kantrowitz, Arthur. "Proposal for an Institution of Scientific Judgment." *Science* 156, no. 3776 (1967): 763–64.

———. "The Test: Meeting the Challenge of New Technology." *Bulletin of the Atomic Scientists* 25, no. 9 (1969): 20–22, 48.

———. "Controlling Technology Democratically." *American Scientist* 63 (1975): 505–9.

———. "The Science Court Experiment." *Jurimetrics Journal* 17 (1977): 332–41.

Kasperson, Roger E., Ortwin Renn, Paul Slovic, Halina S. Brown, Jacque Emel, Robert Goble, Jeanne X. Kasperson, and Samuel Ratick. "The Social Amplification of Risk: A Conceptual Framework." *Risk Analysis* 8, no. 2 (1988): 178–87.

Keller, Evelyn Fox. *A Feeling for the Organism: The Life and Work of Barbara McClintock.* San Francisco: W. H. Freeman and Company, 1983.

Kelly, Kevin. *What Technology Wants.* New York: Penguin, 2010.

Keppo, Jussi, Marianna Kudlyak, Elena Quercioli, Lones Smith, and Andrea Wilson. "For Whom the Bell Tolls: Avoidance Behavior at Breakout in COVID19." Technical report, working paper, 2020.

Kevles, Daniel J. *The Physicists: The History of a Scientific Community in Modern America.* New York: Knopf, 1995. First published 1978.

King, Desmond. "Creating a Funding Regime for Social Research in Britain: The Heyworth Committee on Social Studies and the Founding of the Social Science Research Council." *Minerva* 35, no. 1 (1997): 1–26.

Kirkland, Anna. *Vaccine Court: The Law and Politics of Injury.* New York: NYU Press, 2016.

Kitcher, Philip. *Science, Truth, and Democracy.* Oxford: Oxford University Press, 2001.

Kleinman, Daniel L. *Politics on the Endless Frontier: Postwar Research Policy in the United States.* Durham, NC: Duke University Press, 1995.

Kline, Wendy. *Bodies of Knowledge: Sexuality, Reproduction, and Women's Health in the Second Wave.* Chicago: University of Chicago Press, 2010.

Knight, Jack, and James Johnson. "What Sort of Equality Does Deliberative Democracy Require?" In *Deliberative Democracy: Essays on Reason and Politics,* edited by James Bohman and William Rehg, 279–321. Cambridge, MA: MIT Press, 1997.

Kolodny, Niko. "Rule over None I: What Justifies Democracy?" *Philosophy and Public Affairs* 42 (2014): 287–336.

Krause, Sharon. *Freedom beyond Sovereignty: Reconstructing Liberal Individualism.* Chicago: University of Chicago Press, 2015.

Kuhn, Thomas. *The Structure of Scientific Revolutions.* Chicago: University of Chicago Press, 2012. First published 1962.

Lacey, Hugh. "Rehabilitating Neutrality." *Philosophical Studies* 163, no. 1 (2013): 77–83.

———. "A View of Scientific Methodology as a Source of Ignorance in Controversies about Genetically Engineered Crops." In *Science and the Production of Ignorance: When the Quest for Knowledge Is Thwarted,* edited by Janet Kourany and Martin Carrier, 245–70. Cambridge, MA: MIT Press, 2020.

Lafont, Cristina. "Can Democracy Be Deliberative and Participatory? The Democratic Case for Political Uses of Mini-Publics." *Daedalus* 146, no. 3 (2017): 85–105.

Lamb, Michael, and Melissa Lane. "Aristotle on the Ethics of Communicating Climate Change." In *Climate Justice in a Non-Ideal World,* edited by Clare Heyward and Dominic Rose, 229–55. Oxford: Oxford University Press, 2016.

Landauer, Matthew. "The *Idiōtēs* and the Tyrant: Two Faces of Unaccountability in Democratic Athens." *Political Theory* 42, no. 2 (April 2012): 139–66.

Landemore, Hélène. *Democratic Reason: Politics, Collective Intelligence, and the Rule of the Many.* Princeton, NJ: Princeton University Press, 2008.

Lander, Eric S., Françoise Baylis, Feng Zhang, Emmanuelle Charpentier, Paul Berg, Catherine Bourgain, Bärbel Friedrich, et al., "Adopt a Moratorium on Heritable Genome Editing." *Nature* 567, no. 7747 (2019): 165–68.

Lane, Melissa. "When the Experts Are Uncertain: Scientific Knowledge and the Ethics of Democratic Judgment." *Episteme* 11, no. 1 (March 2014): 97–118.

Lang, Amy. "Agenda-Setting in Deliberative Forums." In *Designing Deliberative Democracy: The British Columbia Citizens' Assembly,* edited by Mark E. Warren and Hilary Pearse, 85–105. Cambridge: Cambridge University Press, 2008.

Larsen, Otto N. *Milestones and Millstones: Social Science at the National Science Foundation, 1945–1991.* New Brunswick, NJ: Transaction Publishers, 1992.

Latour, Bruno. *Politics of Nature: How to Bring the Sciences into Democracy.* Cambridge, MA: Harvard University Press, 2004.

Lawford-Smith, Holly. "Understanding Political Feasibility." *Journal of Political Philosophy* 21, no. 3 (2013): 243–59.

Lempert, Richard. "Civil Juries and Complex Cases: Taking Stock after Twelve Years." In *Verdict: Assessing the Civil Jury System,* edited by Robert Litan, 181–247. Washington, DC: Brookings Institution, 1993.

Lepkowski, Wil. "USA: Science Court on Guard." *Nature* 263 (1976): 454–55.

Levine, Robert J. *Ethics and Regulation of Clinical Research.* Baltimore: Urban and Schwarzenberg, 1986.

Lipsitch, Marc. "Why Do Exceptionally Dangerous Gain-of-Function Experiments in Influenza?" In *Influenza Virus,* edited by Yohei Yamauchi, 589–608. New York: Humana Press, 2018.

———. "Good Science Is Good Science." *Boston Review,* May 12, 2020. http://bostonreview.net /science-nature/marc-lipsitch-good-science-good-science.

Lloyd, Elisabeth. *The Case of the Female Orgasm: Bias in the Science of Evolution.* Cambridge, MA: Harvard University Press, 2009.

Longino, Helen. *Science as Social Knowledge: Values and Objectivity in Scientific Inquiry.* Princeton, NJ: Princeton University Press, 1990.

Longo, Matthew, and Bernardo Zacka. "Political Theory in an Ethnographic Key." *American Political Science Review* 113, no. 4 (2019): 1066–70.

Lukes, Steven. *Power: A Radical View*. London: Macmillan, 1974.

Mahony, Martin, and Mike Hulme. "Modelling and the Nation: Institutionalising Climate Prediction in the UK, 1988–92." *Minerva* 54, no. 4 (2016): 445–70.

Manin, Bernard. "Political Deliberation and the Adversarial Principle." *Daedalus* 146, no. 3 (Summer 2017): 28–38.

Mann, Alfred K. *For Better or for Worse: The Marriage of Science and Government in the United States*. New York: Columbia University Press, 2000.

Mansbridge, Jane, James Bohman, Simone Chambers, Thomas Christiano, Archon Fung, John Parkinson, Dennis F. Thompson, and Mark E. Warren. "A Systemic Approach to Deliberative Democracy." In *Deliberative Systems: Deliberative Democracy at the Large Scale*, edited by John Parkinson and Jane Mansbridge, 1–26. Cambridge: Cambridge University Press, 2012.

Mansbridge, Jane, James Bohman, Simone Chambers, David Estlund, Andreas Føllesdal, Archon Fung, Cristina Lafont, Bernard Manin, and José Luis Martínez. "The Place of Self-Interest and the Role of Power in Deliberative Democracy." *Journal of Political Philosophy* 18, no. 1 (2010): 64–100.

Mazzucato, Mariana. *The Entrepreneurial State*. London: Demos, 2011.

McGarity, Thomas, and Wendy Wagner. *Bending Science: How Special Interests Corrupt Public Health Research*. Cambridge, MA: Harvard University Press, 2008.

McKinnon, Catriona. "Sleepwalking into Lock-in? Avoiding Wrongs to Future People in the Governance of Solar Radiation Management Research." *Environmental Politics* 28, no. 3 (2019): 441–59.

McNeil, Donald G., Jr. "White House to Cut Funding for Risky Biological Study." *New York Times*, October 17, 2014.

———. "A Federal Ban on Making Lethal Viruses Is Lifted." *New York Times*, December 19, 2017. https://www.nytimes.com/2017/12/19/health/lethal-viruses-nih.html.

McQueen, Alison. *Political Realism in Apocalyptic Times*. Cambridge: Cambridge University Press, 2017.

———. "Salutary Fear? Hans Morgenthau and the Politics of Existential Crisis." *American Political Thought* 6, no. 1 (2017): 78–105.

Michaels, David. "Manufactured Uncertainty: Contested Science and the Protection of the Public's Health and Environment." In *Agnotology: The Making and Unmaking of Ignorance*, edited by Robert Proctor and Londa Schiebinger, 90–108. Stanford, CA: Stanford University Press, 2008.

Miller, David. "Justice, Democracy and Public Goods." In *Justice and Democracy: Essays for Brian Barry*, edited by Keith Dowding, Robert E. Goodin, and Carole Pateman, 127–49. Cambridge: Cambridge University Press, 2004.

Mills, Charles. "'Ideal Theory' as Ideology." *Hypatia* 20, no. 3 (2005): 165–84.

Mirowski, Philip. *Science-Mart: Privatizing American Science*. Cambridge, MA: Harvard University Press, 2011.

———. "The Future(s) of Open Science." *Social Studies of Science* 48, no. 2 (2018): 171–203.

Moore, Alfred. "Beyond Participation: Opening Up Political Theory in STS." *Social Studies of Science* 40, no. 5 (2010): 793–99.

———. *Critical Elitism: Deliberation, Democracy, and the Problem of Expertise*. Cambridge: Cambridge University Press, 2017.

Morrow, David R., Robert E. Kopp, and Michael Oppenheimer. "Toward Ethical Norms and Institutions for Climate Engineering Research." *Environmental Research Letters* 4, no. 4 (2009): 1–8.

Morton, Oliver. *The Planet Remade: How Geoengineering Could Change the World.* Princeton, NJ: Princeton University Press, 2016.

Munoz-Dardé, Veronique. "In the Face of Austerity: The Puzzle of Museums and Universities." *Journal of Political Philosophy* 21, no. 2 (2013): 221–42.

Näsström, Sofia. "The Challenge of the All-Affected Principle." *Political Studies* 59, no. 1 (2011): 116–34.

National Commission for the Protection of Human Subjects of Biomedical and Behavioral Research. "The Belmont Report: Ethical Principles and Guidelines for the Protection of Human Subjects of Research." US Department of Health and Human Services, April 18, 1979.

Neal, Homer A., Tobin L. Smith, and Jennifer B. McCormick, *Beyond Sputnik: U.S. Science Policy in the Twenty-First Century.* Ann Arbor: University of Michigan Press, 2008.

Neustadt, Richard E., and Harvey V. Fineberg. *The Swine Flu Affair: Decision Making on a Slippery Disease.* Washington, DC: National Academies Press, 1978.

Ngumbi, Esther. "How to Tackle Repetitive Droughts in the Horn of Africa." *Al Jazeera*, February 14, 2017. https://www.aljazeera.com/indepth/opinion/2017/02/tackle-repetitive -droughts-horn-africa-170214090108648.html.

Nicholson, Joshua M., and John P. A. Ioannidis. "Conform and Be Funded." *Nature* 492 (2012): 34–36.

"Nobel Laureates and Science Groups Demand NIH Review Decision to Kill Coronavirus Grant." *Science News*, May 21, 2020.

Nosengo, Nicola. "Italian Court Finds Seismologists Guilty of Manslaughter." *Nature*, October 22, 2012. https://www.nature.com/news/italian-court-finds-seismologists-guilty-of -manslaughter-1.11640.

Nozick, Robert. *Anarchy, State, and Utopia.* New York: Harper and Row, 1974.

Nussbaum, Martha. *Upheavals of Thought: The Intelligence of the Emotions.* Cambridge: Cambridge University Press, 2001.

———. *Not for Profit: Why Democracy Needs the Humanities.* Princeton, NJ: Public Square, 2016.

Ober, Josiah. *Democracy and Knowledge: Innovation and Learning in Classical Athens.* Princeton, NJ: Princeton University Press, 2008.

O'Neill, John. "The Rhetoric of Deliberation: Some Problems in Kantian Theories of Deliberative Democracy." *Res Publica* 8 (2002): 249–68.

"Open Letter Advocating for an Anti-Racist Public Health Response." June 5, 2020. https://www .calvoices.org/post/open-letter-advocating-for-an-anti-racist-public-health-response-to -demonstrations.

"An Open Letter to the United Nations Convention on Certain Conventional Weapons." https://futureoflife.org/autonomous-weapons-open-letter-2017.

Oppenheimer, Michael, Naomi Oreskes, Dale Jamieson, Keynyn Brysse, Jessica O'Reilly, Matthew Shindell, and Milena Wazeck. *Discerning Experts: The Practices of Scientific Assessment for Environmental Policy.* Chicago: University of Chicago Press, 2019.

Oreskes, Naomi, and Eric Conway. *Merchants of Doubt: How a Handful of Scientists Obscured the Truth on Issues from Tobacco Smoke to Global Warming.* New York: Bloomsbury Press, 2010.

Oreskes, Naomi, Kristin Shrader-Frechette, and Kenneth Belitz. "Verification, Validation, and Confirmation of Numerical Models in the Earth Sciences." *Science* 263, no. 5147 (February 1994): 641–46.

Owens, Susan. *Knowledge, Policy, and Expertise: The UK Royal Commission on Environmental Pollution 1970–2011.* Oxford: Oxford University Press, 2015.

Page, Scott. *Diversity and Complexity.* Princeton, NJ: Princeton University Press, 2010.

Pareek, Manish, Mansoor N. Bangash, Nilesh Pareek, Daniel Pan, Shirley Sze, Jatinder S. Minhas, Wasim Hanif, and Kamlesh Khunti. "Ethnicity and COVID-19: An Urgent Public Health Research Priority." *Lancet* 395, no. 10234 (May 2020): 1421–22.

Parker, Wendy. "Confirmation and Adequacy-for-a-Purpose in Climate Modelling." *Proceedings of the Aristotelian Society Supplementary Volumes* 83, no. 1 (2009): 233–49.

Parker, Wendy, and Eric Winsberg. "Values and Evidence: How Models Make a Difference." *European Journal for Philosophy of Science* 8, no. 1 (2018): 125–42.

Parson, Edward, and Lia Ernst. "International Governance of Climate Engineering." *Theoretical Inquiries in Law* 14, no. 1 (2013): 307–38.

Parson, Edward, and David Keith. "End the Deadlock on Governance of Geoengineering Research." *Science* 339, no. 6125 (2013): 1278–79.

Perceptions of Science in America: A Report from the Public Face of Science Initiative. Cambridge, MA: American Academy of Arts and Sciences, 2018.

Pettit, Philip. "Representation, Responsive and Indicative." *Constellations* 17, no. 3 (2010): 426–34.

———. *On the People's Terms: A Republican Theory and Model of Democracy.* Cambridge: Cambridge University Press, 2012.

Pidgeon, Nick, Roger E. Kasperson, and Paul Slovic, eds. *The Social Amplification of Risk.* Cambridge: Cambridge University Press, 2003.

Pielke, Roger A., Jr. *The Honest Broker: Making Sense of Science in Policy and Politics.* Cambridge: Cambridge University Press, 2007.

Plutynski, Anya. "Safe or Sorry? Cancer Screening and Inductive Risk." In *Exploring Inductive Risk: Case Studies of Values in Science,* edited by Kevin Elliott and Ted Richards, 149–70. New York: Oxford University Press, 2017.

Polanyi, Michael. "Cultural Significance of Science." *Nature* 147 (1941): 119.

———. "The Growth of Thought in Society." *Economica* 8 (1941): 428–56.

———. "The Autonomy of Science." *Scientific Monthly* 60 (1945): 141–50.

———. "The Planning of Science." *Political Quarterly* 16 (1945): 316–28.

———. *The Logic of Liberty: Reflections and Rejoinders.* Chicago: University of Chicago Press, 1951.

———. "The Republic of Science: Its Political and Economic Theory." *Minerva* 1 (1962): 54–73.

Polovedo, Elisabetta, and Henry Fountain. "Italy Orders Jail Terms for 7 Who Did Not Warn of Earthquake." *New York Times,* October 22, 2012. https://www.nytimes.com/2012/10/23/world/europe/italy-convicts-7-for-failure-to-warn-of-quake.html.

Post, Robert. "The Supreme Court Opinion as Institutional Practice: Dissent, Legal Scholarship, and Decisionmaking in the Taft Court." *Minnesota Law Review* 85 (2000): 1267–391.

Powell, Maria, and Mathilde Colin. "Meaningful Citizen Engagement in Science and Technology." *Science Communication* 30, no. 1 (2008): 126–36.

Proctor, Robert N. *Value-Free Science? Purity and Power in Modern Knowledge*. Cambridge, MA: Harvard University Press, 1991.

Proctor, Robert N., and Londa Schiebinger, eds. *Agnotology: The Making and Unmaking of Ignorance*. Stanford, CA: Stanford University Press, 2008.

Putnam, Hilary. *The Collapse of the Fact/Value Dichotomy*. Cambridge, MA: Harvard University Press, 2002.

Qiu, Jane. "How China's 'Bat Woman' Hunted Down Viruses from SARS to the New Coronavirus." *Scientific American*, June 1, 2020.

Rahman, Sabeel. *Democracy against Domination*. New York: Oxford University Press, 2016.

Ravetz, Jerome R. "Usable Knowledge, Usable Ignorance: Incomplete Science with Policy Implications." *Knowledge* 9, no. 1 (1987): 87–116.

Rawls, John. "Two Concepts of Rules." *Philosophical Review* 64, no. 1 (January 1955): 3–32.

———. *A Theory of Justice*. Cambridge, MA: Belknap Press of Harvard University Press, 1971.

———. *Political Liberalism*. New York: Columbia University Press, 1996.

———. *Justice as Fairness: A Restatement*. Cambridge, MA: Harvard University Press, 2001.

Rayner, Steve. "Cultural Theory and Risk Analysis." In *Social Theories of Risk*, edited by Sheldon Krimsky and Dominic Golding, 83–115. Westport, CT: Praeger, 1992.

Raz, Joseph. *The Morality of Freedom*. Oxford: Clarendon Press, 1986.

"Read the State Department Cable That Launched Claims That Coronavirus Escaped from Chinese Lab." *Washington Post*, July 17, 2020.

Reiss, Julian. "A Pragmatist Theory of Evidence." *Philosophy of Science* 82, no. 3 (2015): 341–62.

Renn, Ortwin. *Risk Governance: Coping with Uncertainty in a Complex World*. London: Earthscan, 2008.

Resnik, David. "Dual-Use Research and Inductive Risk." In *Exploring Inductive Risk: Case Studies of Values in Science*, edited by Kevin Elliott and Ted Richards, 59–78. New York: Oxford University Press, 2017.

Rice, Doyle, and Ledyard King. "Trump's Budget Proposal 'Savages' Climate Research, Scientists Say." *USA Today*, May 23, 2017. https://www.usatoday.com/story/news/nation/2017/05/23/trumps-budget-proposal-savages-climate-research-scientists-say/102062556/.

Richardson, Henry S. *Democratic Autonomy: Public Reasoning about the Ends of Policy*. New York: Oxford University Press, 2002.

Robock, Alan. "20 Reasons Why Geoengineering May Be a Bad Idea." *Bulletin of the Atomic Scientists* 64, no. 2 (2008): 14–18.

———. "Is Geoengineering Research Ethical?" *Security and Peace* 30, no. 4 (2012): 226–29.

Robock, Alan, Martin Bunzl, Ben Kravitz, and Georgiy L. Stenchikov. "A Test for Geoengineering?" *Science* 327, no. 5965 (2010): 530–31.

Rosenblum, Nancy L. *On the Side of the Angels: An Appreciation of Parties and Partisansip*. Princeton, NJ: Princeton University Press, 2008.

———. "Governing beyond Imagination: The 'World Historical' Sources of Democratic Dysfunction." *Boston University Law Review* 94 (2014): 649–69.

Rosenblum, Nancy L., and Russell Muirhead. *A Lot of People Are Saying: The New Conspiracism and the Assault on Democracy*. Princeton, NJ: Princeton University Press, 2019.

Rosanvallon, Pierre. *Democratic Legitimacy: Impartiality, Reflexivity, Proximity*. Translated by Arthur Goldhammer. Princeton, NJ: Princeton University Press, 2010.

Ross, Jacob. "Rejecting Ethical Deflationism." *Ethics* 116, no. 4 (2006): 742–68.

Royal Society. *Geoengineering the Climate: Science Governance and Uncertainty.* Policy document 10/09, 2009.

Rudner, Richard. "The Scientist qua Scientist Makes Value Judgments." *Philosophy of Science* 20, no. 1 (January 1953): 1–6.

Sabel, Charles, Archon Fung, and Bradley Karkkainen. *Beyond Backyard Environmentalism.* Boston: Beacon Press, 2000.

Salkin, Wendy. *"Informal Political Representation: Normative and Conceptual Foundations."* PhD diss., Harvard University, 2018.

Sanders, Lynn. "Against Deliberation." *Political Theory* 25, no. 3 (1997): 347–76.

Sarewitz, Daniel. "How Science Makes Environmental Controversies Worse." *Environmental Science and Policy* 7 (2004): 385–403.

———. "Normal Science and Limits on Knowledge: What We Seek to Know, What We Choose Not to Know, What We Don't Bother Knowing." *Social Research* 77, no. 3 (2010): 997–1010.

Satkunanandan, Shalini. "Max Weber and the Ethos of Politics beyond Calculation," *American Political Science Review* 108, no. 4 (2014): 169–81.

Savage, James D. *Funding Science in America: Congress, Universities, and the Politics of the Academic Pork Barrel.* Cambridge: Cambridge University Press, 1999.

Sax, Sarah. "That Geoengineering Gender Problem." *Canada's National Observer*, December 23, 2019. https://www.nationalobserver.com/2019/12/23/news/geoengineering-gender -problem.

Schattschneider, Elmer E. *The Semi-Sovereign People.* New York: Holt, Rinehalt and Winston, 1960.

Schmitt, Michael, and Jeffrey Thurnher. "Out of the Loop: Autonomous Weapon Systems and the Law of Armed Conflict." *Harvard National Security Journal* 4, no. 2 (2013): 231–81.

Scholnick, David. "How a $47 Shrimp Treadmill Became a $3-Million Political Plaything." *Chronicle of Higher Education*," November 13, 2014. https://www.chronicle.com/blogs/conversation /2014/11/13/how-a-47-shrimp-treadmill-became-a-3-million-political-plaything.

Schroeder, S. Andrew. "Using Democratic Values in Science: An Objection and (Partial) Response." *Philosophy of Science* 84, no. 5 (2017): 1044–54.

Shuman, Daniel W., Elizabeth Whitaker, and Anthony Champagne, "An Empirical Examination of the Use of Expert Witnesses in the Courts—Part II: A Three City Study." *Jurimetrics* 34, no. 2 (Winter 1994): 193–208.

Simpson, Thomas, and Vincent Müller. "Just War and Robots' Killings." *Philosophical Quarterly* 66, no. 263 (2016): 302–22.

Sinden, Amy. "In Defense of Absolutes: Combating the Politics of Power in Environmental Law." *Iowa Law Review* 90 (2004–5): 1405–513.

Sloan, Frank A., Penny B. Githens, Ellen Wright Clayton, Gerald B. Hickson, Douglas A. Gentile, and David F. Partlett, *Suing for Medical Malpractice.* Chicago: University of Chicago Press, 1993.

Slovic, Paul. "Perceived Risk, Trust, and Democracy." *Risk Analysis* 13, no. 6 (1993): 675–82.

Slovic, Paul, Baruch Fischhoff, and Sarah Lichtenstein. "Facts and Fears: Understanding Perceived Risk." In *Societal Risk Assessment: How Safe Is Safe Enough?*, edited by Richard C. Schwing and Walter A. Albers, 181–214. New York: Plenum Press.

Snyder, Laura. *Reforming Philosophy: A Victorian Debate on Science and Society*. Chicago: University of Chicago Press, 2006.

Solomon, Miriam. "Standpoint and Creativity." *Hypatia* 24, no. 4 (2009): 226–37.

———. *Making Medical Knowledge*. New York: Oxford University Press, 2015.

Somin, Ilya. "Voter Ignorance and the Democratic Ideal." *Critical Review* 12, no. 4 (1998): 413–58.

Sparrow, Robert. "Killer Robots." *Journal of Applied Philosophy* 24, no. 1 (2007): 62–77.

Steel, Daniel. *Philosophy and the Precautionary Principle: Science, Evidence, and Environmental Policy*. Cambridge: Cambridge University Press, 2015.

Steele, Katie. "The Scientist qua Policy Advisor Makes Value Judgments." *Philosophy of Science* 79, no. 5 (2012): 893–904.

Taragin, Mark I., Laura R. Willett, Adam P. Wilczek, Richard Trout, and Jeffrey L. Carson. "The Influence of Standard of Care and Severity of Injury on the Resolution of Medical Malpractice Claims." *Annals of Internal Medicine* 117, no. 9 (1992): 780–84.

Taylor, Isaac. "Who Is Responsible for Killer Robots? Autonomous Weapons, Group Agency, and the Military-Industrial Complex." *Journal of Applied Philosophy* (2020). doi.org/10.1111/japp.12469.

Thompson, Dennis. "Who Should Govern Who Governs? The Role of Citizens in Reforming the Electoral System." In *Designing Deliberative Democracy: The British Columbia Citizens' Assembly*, edited by Mark E. Warren and Hilary Pearse, 20–49. Cambridge: Cambridge University Press, 2008.

Thomson, Judith Jarvis, Catherine Elgin, David A. Hyman, Philip E. Rubin, and Jonathan Knight. "Research on Human Subjects: Academic Freedom and the Institutional Review Board." *Academe* 92, no. 5 (2006): 95–100.

Tollefson, Jeff. "First Sun-Dimming Experiment Will Test a Way to Cool the Earth." *Nature* 563, no. 7733 (2018): 613–15.

Toxvaerd, Flavio. "Equilibrium Social Distancing." Cambridge-INET Working Paper Series No: 2020/08, March 2020. http://www.econ.cam.ac.uk/research-files/repec/cam/pdf/cwpe2021.pdf.

Tuck, Richard. *Free Riding*. Cambridge, MA: Harvard University Press, 2008.

Tucker, Aviezer. "Pre-Emptive Democracy: Oligarchic Tendencies in Deliberative Democracy." *Political Studies* 56, no. 1 (2008): 127–47.

Turner, Stephen. "What Is the Problem with Experts?" *Social Studies of Science* 31, no. 1 (2001): 123–49.

———. *Liberal Democracy 3.0: Civil Society in an Age of Experts*. London: Sage, 2003.

Urfalino, Philippe. "Reasons and Preferences in Medicine Evaluation Committees." In *Collective Wisdom: Principles and Mechanisms*, edited by Hélène Landemore and Jon Elster, 173–202. Cambridge: Cambridge University Press, 2012.

US National Academy of Sciences. "About Our Expert Consensus Reports." http://dels.nas.edu/global/Consensus-Report.

———. "Mission." http://www.nasonline.org/about-nas/mission/.

———. "What We Do." http://www.nationalacademies.org/about/whatwedo/index.html.

van der Bles, Anne Marthe, Sander van der Linden, Alexandra L. J. Freeman, and David J. Spiegelhalter. "The Effects of Communicating Uncertainty on Public Trust in Facts and Numbers." *Proceedings of the National Academy of Sciences* 117, no. 14 (2020): 7672–83.

Vidmar, Neil. "The Performance of the American Civil Jury." *Arizona Law Review* 40 (1989): 849–901.

Vidmar, Neil, and Shari Seidman Diamond. "Juries and Expert Evidence." *Brooklyn Law Review* 66 (2001): 1121–81.

Vidmar, Neil, Shari Seidman Diamond, Mary R. Rose, and René Stemple Ellis. "Juror Discussions during Civil Trials: Studying an Arizona Innovation." *Arizona Law Review* 45 (2003): 1–83.

Waldman, Scott. "Trump Administration Seeks Big Budget Cuts for Climate Research. *Scientific American*, March 7, 2017. https://www.scientificamerican.com/article/trump-administration -seeks-big-budget-cuts-for-climate-research.

Warren, Mark E. "Democratic Theory and Trust." In *Democracy and Trust*, edited by Mark E. Warren, 310–45. Cambridge: Cambridge University Press, 1999.

———. "Governance Driven Democratization." *Critical Policy Studies* 3, no. 1 (2009): 3–13.

Warren, Mark E., and John Gastil. "Can Deliberative Minipublics Address the Cognitive Challenges of Democratic Citizenship?" *Journal of Politics* 77, no. 2 (2015): 562–74.

Warren, Mark E., and Hilary Pearse, eds. *Designing Deliberative Democracy: The British Columbia Citizens' Assembly.* Cambridge: Cambridge University Press, 2008.

"Was Duck Penis Study Appropriate Use of Taxpayer Money?" Fox News, March 25, 2013. http://www.foxnews.com/opinion/2013/03/25/was-duck-penis-study-appropriate-use -taxpayer-money.html.

Weber, Max. "Bureaucracy." In *From Max Weber: Essays in Sociology*, edited by Hans Heinrich Gerth and C. Wright Mills, 196–244. New York: Oxford University Press, 1958.

Weingart, Peter. "Scientific Expertise and Political Accountability: Paradoxes of Science in Politics." *Science and Public Policy* 26, no. 3 (1999): 151–61.

Wicksell, Knut. "A New Principle of Just Taxation." In *Classics in the Theory of Public Finance*, edited by Richard Musgrave and Alan Peacock, 72–118. London: Palgrave Macmillan, 1958.

Wiens, David. "Prescribing Institutions without Ideal Theory." *Journal of Political Philosophy* 20, no. 1 (2012): 45–70.

Wilford, John Noble. "Science Considers Its Own 'Court.'" *New York Times*, February 29, 1976, 140.

———. "Leaders Endorse Science Court Test." *New York Times,* January 2, 1977, 28.

Williams, Bernard. *Ethics and the Limits of Philosophy*. Abingdon, UK: Routledge, 2006. First published 1985.

Wolff, Jonathan. "Risk, Fear, Blame, Shame and the Regulation of Public Safety." *Economics and Philosophy* 22 (2006): 409–27.

Wolin, Sheldon. "Max Weber: Legitimation, Method, and the Politics of Theory." *Political Theory* 9, no. 3 (1981): 401–24.

———. *Politics and Vision: Continuity and Innovation in Western Political Thought.* Princeton, NJ: Princeton University Press, 2004. First published 1960.

Wolpert, Lewis. "Is Science Dangerous?" In *Scientific Freedom*, edited by Simona Giordano, John Cotton, and Marco Cappato, 31–42. London: Bloomsbury, 2012.

Wootton, David. *Bad Medicine: Doctors Doing Harm since Hippocrates.* New York: Oxford University Press, 2007.

Young, Iris Marion. *Justice and the Politics of Difference*. Princeton, NJ: Princeton University Press, 1990.

———. "Activist Challenges to Deliberative Democracy." *Political Theory*, 29, no. 5 (2001): 670–90.

"Your Tax Dollars at Work: Shrimp on Treadmills." Fox News, May 26, 2011. http://video .foxnews.com/v/960953334001/?#sp=show-clips.

Zacka, Bernardo. *When the State Meets the Street*. Cambridge, MA: Harvard University Press, 2017.

Zimmerman, Michael J. "Intervening Agents and Moral Responsibility." *Philosophical Quarterly* 35, no. 141 (1985): 347–58.

INDEX

accountability: during the COVID-19 pandemic, 198, 201; distrust as an underlying mechanism of, 102; electoral politics and, 91–92; free public discourse as a requirement for democratic, 151; of jurors, 123; minipublics and, 104, 107; publicly-funded science and democratic, 151–52, 157, 160; scientific advisory bodies and, 84–85

acid rain, 34

advice: expert (*see* experts/expertise); paradox of scientific, 64–65; requirements of for policy decisions, 64; scientific advisory bodies (*see* scientific advisory bodies)

agenda setting: by organizers in minipublics, 105; research funding decisions and, 134, 152–55; role of science in, 47–48; in the science court, 112–15

agroecology movement, 37, 50

AIDS crisis, 45–46, 58–59

"all-subjected principle," 6n10

American Association for the Advancement of Science, 3, 99

Arizona Jury Project, 130–31

asymmetrical knowledge: in deliberation, 108–10; public debate under conditions of, question of, 84; science court and, 124

authority: advisers hiding disagreement to bolster, 95; appeals to scientific, 4–5, 206; deliberation under conditions of unequal, 108–11; of jurors in the science court, 123; relationship between scientific and political, 5, 60, 67, 186; of the science court, 123–25, 127; of scientists, 18, 64–65

Barry, Brian, 145–46

Belmont report, 167, 172

Bentham, Jeremy, 188

Berlin, Isaiah, 9

Betz, Gregor, 75, 77

Biden, Joseph, 199

Black Lives Matter protests, 198

Bonaparte, Napoléon, 22, 97

Bostrom, Nick, 164n6

Bratman, Michael, 78

British Columbia Citizens' Assembly, 107

Brown, Mark B., 11n24, 17, 82–83

bureaucracy/bureaucrats: funding decisions by, opposition to, 159; role of in 1976 swine flu outbreak, 92–93; scientists and, difference between, 18–19, 50–51, 66–67; Weber on, 7, 9, 10n21, 185

Bush, Vannevar, 135–39, 143, 147

Carter, Jimmy, 100

Cartwright, Nancy, 31

Centers for Disease Control (CDC), swine flu outbreak of 1976, 92–93

Chan, Alina, 208–9

Christiano, Thomas, 101–2

climate change: geoengineering to counteract the effects of, 164–65 (*see also* stratospheric aerosol injection (SAI)); modeling choices in research on, 40–41; Trump administration attack on research related to, 133–34; two-sides journalism on, 89; uncertainty and disagreement among scientists regarding, 20–21

Coen, Deborah, 2n8

funding for science; scientific autonomy versus democratic control); goals and process of, 55–56; identifying and using corrective information as second stage of, 57; inadequacy of transparency and communication from scientists as an alternative to, 58–60; nonexperts, requirements placed on, 85; novelty of the proposal for, 61–62; objections to the proposal for, 90–95; science court as a contribution to, 124 (*see also* new science court; science court); scientific advice reconceptualized as a facilitator of public debate, 84–85; small-scale democratic experiments/minipublics as a solution for (*see* small-scale democratic experiments/ minipublics); values in science, as the democratic antidote for, 61

Dewey, John, 7, 151, 187

dissenting opinions: during the COVID-19 pandemic, 199–203, 207–9; Millian function of, 127, 209; research funding and, 156–60; from scientific advisory bodies, 86–95; unreasonable views and, 120–21, 209–10; from U.S. Supreme Court decisions, 85–86

division of labor models: of elite-nonelite relationship, 101–3; of expert-laypeople relationship, 8–10, 50–52, 67–68; redrawing the Weberian, goal of, 11–12

domination by experts, 18, 48–49

Douglas, Heather, 15–16, 30n7, 42, 53–55, 69

Dupré, John, 31, 33

Einstein, Albert, 142

elites/elitism: Bush's vision of public funding, 126–27; elite-nonelite division of labor, 101–3; elite power to guide public imaginary, 182–83; Kuhn's view of the scientific community, 136; rejecting elitist views of democracy, 17, 22, 51–52

Epstein, Steven, 27–28, 37

equality: bureaucratic delegation of power to experts as threat to, 50–51; deliberation and, 108–9; democracy characterized by, 21; the democratic problem with the role of scientists and, 50

ethical scientist, the, 80–83

European Union, neutrality of medical licensing committees, 76

Evans, Robert, 8n16, 76n29

evidence: admissibility of in new science court, 118–21; competing approaches to during the COVID-19 pandemic, 42–44; determination of good quality, standards for, 41–45; mismatch between scientific and political purposes in assessing, 42; randomized control trials as the gold standard for, 43–44 (*see also* randomized control trials)

experts/expertise: democracy/citizens and, relationship between (*see* science/expertise-democracy/citizens relationship); "expert," definitions of, 8n16; laypeople/the public and, division of labor model regarding, 8–10, 50–51, 67–68; neutrality as the basis for credibility of, 64–65 (*see also* neutrality); political representation and, 82–83 (*see also* representation); rule by, 6; scientific advisory bodies (*see* scientific advisory bodies); value judgments, shaped by, 60

Ezrahi, Yaron, 55n45

feasibility, determination of, 48

Federal Communications Commission, 88n43

feminist philosophy/philosophers, 35–36

Feyerabend, Paul, 142

Fishkin, James, 106

Ford, Gerald, 92, 99–100

Fox News, 133

freedom: of discussion as inadequate for critical scrutiny, 115; of inquiry, 163, 173; limiting to prevent wrongdoing, permissibility of, 169–71

Fuller, Jonathan, 42–43

GPSR Authorized Representative: Easy Access System Europe - Mustamäe tee 50, 10621 Tallinn, Estonia, gpsr.requests@easproject.com

www.ingramcontent.com/pod-product-compliance
Ingram Content Group UK Ltd.
Pitfield, Milton Keynes, MK11 3LW, UK
UKHW040205190325
456107UK00005B/134